CITIES AT RISK

Global Urban Studies

Series Editor:
Laura A. Reese, Michigan State University, USA

Providing cutting-edge interdisciplinary research on spatial, political, cultural and economic processes and issues in urban areas across the US and the world, volumes in this series examine the global processes that impact and unite urban areas. The organizing theme of the book series is the reality that behavior within and between cities and urban regions must be understood in a larger domestic and international context. An explicitly comparative approach to understanding urban issues and problems allows scholars and students to consider and analyze new ways in which urban areas across different societies and within the same society interact with each other and address a common set of challenges or issues.

Books in the series cover topics which are common to urban areas globally, yet illustrate the similarities and differences in conditions, approaches, and solutions across the world, such as environment/brownfields, sustainability, health, economic development, culture, governance and national security. In short, the *Global Urban Studies* book series takes an interdisciplinary approach to emergent urban issues using a global or comparative perspective.

Cities at Risk

Planning for and Recovering from Natural Disasters

Edited by

PIERRE FILION
University of Waterloo, Ontario, Canada

GARY SANDS
Wayne State University, Detroit, USA

MARK SKIDMORE
Michigan State University, Michigan, USA

Routledge
Taylor & Francis Group
LONDON AND NEW YORK

First published 2015 by Ashgate Publishing

Published 2016 by Routledge
2 Park Square, Milton Park, Abingdon, Oxon OX14 4RN
711 Third Avenue, New York, NY 10017, USA

First issued in paperback 2018

Routledge is an imprint of the Taylor & Francis Group, an informa business

British Library Cataloguing in Publication Data
A catalogue record for this book is available from the British Library

The Library of Congress has cataloged the printed edition as follows:
Cities at risk : planning for and recovering from natural disasters / [edited] by Pierre Filion, Gary Sands and Mark Skidmore.
pages cm. -- (Global urban studies)
Includes bibliographical references and index.
ISBN 978-1-4724-4168-3 (hardback)
1. Emergency management. 2. Natural disasters--Planning.
3. Cities and towns. I. Filion, Pierre, 1952- editor. II. Sands, Gary, editor. III. Skidmore, Mark, editor.
HV551.2.C528 2016
363.34'6091732--dc23

2015011052

ISBN 13: 978-1-138-54727-8 (pbk)
ISBN 13: 978-1-4724-4168-3 (hbk)

Contents

	List of Figures	*vii*
	List of Tables	*ix*
	Notes on Contributors	*xi*
1	Introduction: Urban Resilience Efforts in the Face of Natural Disasters *Pierre Filion, Mark Skidmore, and Gary Sands*	1
2	Enhancing Hazard Resilience among Impoverished Urban Communities in Ghana: The Role of Women as Catalysts for Improvement *Robin L. Ersing, Osman Alhassan, Jesse Sey Ayivor, and Kiki Caruson*	15
3	An Analysis of Governmental Performance during Urban Disasters: Fukushima and Hurricane Sandy *Saundra K. Schneider and Marty P. Jordan*	31
4	Urban Disaster Recovery: Considerations for Damage Reduction and Sustainability *Zhila Pooyan*	61
5	The Earthquakes in Lisbon (1755) and Angra do Heroísmo (1980): Reconstruction Processes and Actual Risk *Artur Feio, Roberto Bologna, Daniel Monteiro, and Daniel Félix*	81
6	From Disaster to Opportunity: Reviving Urban Function after the Canterbury Earthquakes *Joanne Stevenson, Sonia Giovinazzi, Erica Seville, Charlotte Brown, Yan (Alice) Chang-Richards, and Suzanne Wilkinson*	97
7	Adaptive Strategies of Urban Disaster Recovery Planning *Mark Kammerbauer*	139

8 Urban Vulnerability and Resiliency to Natural Disasters: An Integrative Tourism Planning Perspective 159
Hyun Kim and David W. Marcouiller

9 Computing the Aggregate Cost of a Disaster: National and Regional Issues 189
Charles Steindel

10 Conclusion: Stages and Scales of Urban Resilience Strategies 205
Gary Sands, Pierre Filion, and Mark Skidmore

Index *215*

List of Figures

2.1	Location of Sabon Zongo and Yeji urban settlements	20
4.1	Permanent public housing, Kobe City	63
4.2	Open space in redevelopment projects, Kobe City	64
4.3	Agra-e-Bam pre-disaster	67
4.4	Abra-e-Bam post-disaster	68
4.5	Tents located near destroyed houses	70
4.6	Temporary prefabricated housing, Bam City	71
4.7	Temporary prefabricated housing, Bam City	72
6.1	Liquefaction silt and sands	98
6.2	Damage to unreinforced masonry	99
6.3	Policy decision makers at various levels of government	101
6.4	Residential property land damage	104
7.1	Lower Ninth Ward, typical shotgun house	151
7.2	Lower Ninth Ward, multifamily development	152
7.3	Lower Ninth Ward, small rental property	153
8.1	Integrative tourism disaster planning model	171

List of Tables

6.1 Categories of land damage 107
6.2 Technical categories for repair and rebuilding 108

7.1 Comparison of recovery planning and master planning 143
7.2 Citations showing federal justification for institutional aspects 148
7.3 Citations showing local critique of institutional aspects 149
7.4 Citations showing impact of Road Home program 149
7.5 Citations showing impact of the Citywide plan 150
7.6 Citations showing impact of Master plan 151

8.1 Literature on community-based disaster response planning by disaster type 165
8.2 Literature related specifically to vulnerability, resiliency, and risk models 176

9.1 New Jersey state revenues 198
9.2 New Jersey reproducible capital, year-end 2011 202

Notes on Contributors

Osman Alhassan, PhD, holds a degree in Geography (Natural Resources and Environment) and is Senior Research Fellow at the Institute of African Studies, University of Ghana. He teaches graduate courses in research methodology, field practice, and disaster risk reduction. As a Senior Fellow at the Centre for Migration Studies, he teaches and researches migration and the African diaspora, and migration and climate change. His research spans 15 years and most recently has focused on urban risks with environmental health and risk accumulation using various neighborhoods in the city of Accra.

Jesse Sey Ayivor, PhD, holds a degree in the Environmental Science Programme, University of Ghana, Legon. He currently works as a Research Fellow with the Institute for Environment and Sanitation Studies, University of Ghana, and has over 10 years of research experience in environmental issues as they relate to land use, local livelihoods and vulnerability. His major research interests are in climate change adaptation and policy, disaster risk management in flood-prone areas, protected area management in relation to local livelihoods, and land tenure as it affects the poor and vulnerable. He also teaches courses in Population and Urbanization Studies.

Roberto Bologna, Architect, PhD in Architectural Technology, Professor of Architectural Technology, Department of Architecture, University of Florence and Director of the master degree in Architecture at the University of Florence. Head of the Department of Architectural Technology and Design "Pierluigi Spadolini" (Dip. TAD) from 2006 to 2012. Member of the Teaching Board of the Research Doctorate in Architecture at the Dept. of Architecture. Enrolled at the Register of the Disciplinary Expert Reviewers for ANVUR (National Agency of University system and Research Evaluation). Member of the CIB W120 commission "Disaster and the built environments." Member of the international research network I-rec "Post disaster reconstruction." Member of the Interuniversity Research Centre TESIS "Systems and Technologies for Health Care Buildings (University of Florence, University of Rome "La Sapienza"). Fields of interest (scientific research and architectural design): university architecture (housing for students, educational research unit), social and health care architecture (hospitals, centers for Alzheimer's patient care), temporary architecture, civil protection (shelters for post disaster emergency, shelter areas).

Charlotte Brown is a research fellow with the Resilient Organizations research program (www.resorgs.org.nz) at the University of Canterbury as well as a tutor in Civil Engineering at the Christchurch Polytechnic Institute of Technology. She has a Bachelor's degree and PhD in Civil Engineering. Charlotte has a wide range of research interests and skills, including: dynamic systems modelling, risk management, policy analysis, solid waste management, environmental management, disaster management and resilience. Her PhD research investigated management of solid waste following disaster events. This research explored technical, social, economic, environmental, regulatory, management and risk implications of solid waste management post-disaster.

Kiki Caruson, PhD, is Associate Professor in the Department of Government and International Affairs, the Assistant Vice President for Research, Innovation and Global Affairs for University of South Florida World, and a Faculty Fellow with USF Research and Innovation. Dr Caruson is an established expert in emergency management, disaster risk reduction, hazard analysis, intergovernmental relations, and US diplomacy. She has authored numerous peer-reviewed publications and is an internationally recognized and engaged scholar. In 2013, Dr Caruson was inducted into the *International Women in Homeland Security and Emergency Management Hall of Fame*.

Yan (Alice) Chang-Richards is a lecturer at the Department of Civil and Environmental Engineering, the University of Auckland, New Zealand. She has a Bachelor (Hons) in Civil Engineering and Master in Management Science and Engineering. Alice obtained her PhD in Civil Engineering at the University of Auckland in 2012, where she investigated and developed a framework for the sourcing of building and engineering resources for reconstruction following large-scale disasters. In 2013, Alice joined the Department of Civil and Environmental Engineering of the University of Auckland to take up the lecturing role in project management and engineering risk management.

Robin L. Ersing, PhD, is Associate Professor in the School of Public Affairs at the University of South Florida. Areas of research include community disaster resilience, the role of women in disaster management planning, coastal resilience and farm worker communities, and community asset mapping. Dr Ersing is PI or Co-PI on several national and international research grants including work in Ghana, Africa. Currently she is part of a six nation grant to study rebuilding livelihoods post-disaster. Among her publications is the book, "Surviving Disaster: The Role of Social Networks" published by Lyceum books in 2012.

Artur Feio is now Assistant Professor at the University Lusíada, Famalicão—Portugal. From 2002 to 2006, he was a Doctoral fellow at the University of Minho, Portugal and University of Florence, Italy where he had performed his PhD work (Inspection and Diagnosis of Historical Timber Structures: NDT Correlations and

Structural Behavior). He has made contributions in the fields of Wood Structures, Structural Rehabilitation of Wood Structures and NDT Tests on Wood Structures. He has had the opportunity to participate in several consultancy projects in Portugal, including inspection diagnosis and strengthening design. He is actively involved in National and European standardization committees.

Daniel Félix is a member of CITAD—Centre of Investigation in Architecture and Design at the University Lusíada, and of HMS—Historical and Masonry Structures at the University of Minho. He is a PhD student at the Faculty of Architecture and Arts of the University Lusíada of Vila Nova de Famalicão, where he obtained an Integrated Master in Architecture. He is conducting his doctorate thesis in post-disaster architecture.

Pierre Filion is a professor at the School of Planning of the University of Waterloo in Canada. His areas of research interest include metropolitan-scale planning, downtowns, suburban areas, and risks facing cities. He has written over 100 articles and chapters and has served on the National Capital Commission (Ottawa) Planning and Real Estate Advisory Committee, the International Joint Commission (for the Great Lakes) Scientific Advisory Committee, and the Ontario Government Central Zone Panel, which has generated the ideas included in the Growth Plan for the Greater Golden Horseshoe, a region focused on Toronto.

Sonia Giovinazzi is a Research Fellow in Risk Management at the Civil and Natural Resources Engineering Department, at University of Canterbury, NZ. She has a Bachelor and Master in Civil Engineering and a PhD in "Risk Management of Natural and Anthropogenic Hazards in the Built Environment." Sonia's current research activities focus on seismic risk analysis and damage scenarios, seismic risk mitigation and management, and post-disaster resilient recovery. Following the Canterbury earthquakes, Sonia led projects funded by the Natural Hazard Research Platform of NZ, to provide on-the-ground support to the initial recovery phase for multiple end-users and agencies.

Marty P. Jordan is a doctoral candidate in the Department of Political Science and recipient of a University Enrichment Fellowship at Michigan State University. His research focuses on state politics and public policy, including disaster policy. He has hands-on experience dealing with urban disasters. While living and working in Central America, he was directly impacted by Hurricane Stan that struck in October 2005. After evacuating his home, Marty served as a liaison between private organizations and governmental agencies and provided emergency relief to affected populations.

Mark Kammerbauer is a German-American urbanist. He holds a diploma in architecture (Technical University of Munich), a Master of Science degree in urban studies (Institute for European Urban Studies, Bauhaus University Weimar) and the title Doktor-Ingenieur (Bauhaus University Weimar). He has held academic positions at the Technical University of Munich and the School of Applied Sciences Nuremberg and is a Visiting Fellow in Planning at the University of Queensland. His teaching responsibilities cover architectural design, urban planning and governance, adaptive planning and urban risk. His research focus is urban recovery after disaster and social vulnerability.

Hyun Kim is a doctoral candidate in Urban and Regional Planning at the University of Wisconsin—Madison with prior graduate training at Pusan National University (South Korea). His academic interests involve the social and economic dimensions of environmental change, environmental planning and natural resource policy for sustainability, natural hazard management, climate change adaptation, community capacity, urban redevelopment and residential mobility, spatial econometrics, and longitudinal data analysis. He has several published works that have appeared in *Tourism Geographies*, *KSCE Journal of Civil Engineering*, *Geographical Journal of Korea*, *Journal of the Korean Society of Civil Engineers*, and *Journal of the Korea Planners Association.*

David W. Marcouiller is Professor of Urban and Regional Planning at the University of Wisconsin—Madison where he serves as Department Chair. He also holds a joint appointment with Cooperative Extension as a development planning content specialist. A resource economist by training, his work focuses on the linkages between natural resources and community economic development with a particular interest in multi-functional rural landscapes, amenity driven migration, and the compatibility of alternative land uses. He has published over 180 manuscripts in a variety of outlets that span tourism and forest economics, outdoor recreation planning, regional science, landscape and urban planning, and rural development. His most recent co-edited book was published by Ashgate and is entitled *Rural Housing, Exurbanization, and Amenity Driven Development.*

Daniel Monteiro, architect, graduated from the University Lusíada of Vila Nova de Famalicão, where he is conducting his master thesis work in the field of post-disaster architecture and pre-disaster planning.

Zhila Pooyan is an assistant professor in the risk management research center of the International Institute of Earthquake Engineering and Seismology in Tehran, Iran. She has a PhD in urban planning from Kobe University, Japan. Her main research interests are disaster recovery, community based disaster management and public awareness of disaster preparedness. She has studied community-based organizations in Kobe city, disaster recovery experiences in Iran including the Manjil (1990) and Bam earthquakes (2003), and also emergency settlement experiences in devastating earthquakes in Iran since 1990.

Gary Sands is Professor Emeritus of Urban Planning at Wayne State University in Detroit, Michigan, where he taught for 36 years. His teaching and research has focused on housing and economic development policies and finance. Sands continues to consult with public agencies and non-profit organizations on these topics. He holds a doctorate in Housing and Public Policy from Cornell University and a Master of Urban Planning degree from Wayne State University.

Saundra K. Schneider is Professor in the Department of Political Science at Michigan State University and Director of the Inter-university Consortium for Political and Social Research Summer Program in Quantitative Methods of Social Research at the University of Michigan. Schneider is the author of *Dealing with Disaster: Public Management in Crisis Situations* (2011) and *Flirting with Disaster: Public Management in Crisis Situations* (1995). Her work has also appeared in a number of leading professional journals. Schneider's current research focuses on the role of administrative forces in public policy-making, particularly in the areas of disaster response, health care, and welfare policy.

Erica Seville is co-leader of the Resilient Organizations research program. She has authored over 70 research articles and is regularly invited to speak on the topic of resilience. Erica is the director of Risk Strategies Research and Consulting, is an Adjunct Senior Fellow with the Department of Civil and Natural Resources Engineering at the University of Canterbury, and sits on the Resilience Expert Advisory Group providing advice and support to the Critical Infrastructure Advisory Committee of the Australian Federal Government. She has a Bachelor of Civil Engineering degree and a PhD in the application of risk assessment techniques.

Mark Skidmore is Professor of Economics at Michigan State University, where he holds the Morris Chair in State and Local Government Finance and Policy. He currently serves as Co-Editor of the *Journal of Urban Affairs*. Professor Skidmore has expertise in public finance and urban/regional economics. Current interests include the economics of natural disasters and the interrelationship between public sector decisions and economic activity. His research has appeared in many professional journals and has been cited in prominent news outlets such as the BBC, Boston Globe, Economist, Forbes, Newsweek, New Yorker, New York Times, PBS News Hour, Reuters, and the Washington Post.

Charles Steindel is Scholar in Residence at the Anisfield School of Business, Ramapo College of New Jersey. From 2010 to 2014, he was Chief Economist of the New Jersey Treasury Department. He came to the Treasury from the Federal Reserve Bank of New York, where he was a Senior Vice President. Prior to working at the New York Fed, he was employed by the First National Bank of Chicago and the Federal Reserve Board. He received his bachelor's degree from

Emory University and his PhD from MIT, and is a member of the Committee on Research in Income and Wealth.

Joanne Stevenson is a research fellow with the Resilient Organisations Research Programme. She obtained her PhD in Geography at the University of Canterbury in 2014, examining place embeddedness and organizational networks following the Canterbury earthquakes. She engages in interdisciplinary research focusing on the nexus of human and environmental systems in hazards and disasters. Her main research interests include social-economic and organizational vulnerability and resilience, and social geographies of disaster recovery. Her work can be found in journals appealing to many disciplines interested in hazards and disasters including: *Earthquake Spectra*; *Natural Hazards Review*; *Disasters*; and *Cartography and Geographic Information Science.*

Suzanne Wilkinson is a professor at the Department of Civil and Environmental Engineering, the University of Auckland. She has a PhD in Construction Management and a Bachelor (Hons) in Civil Engineering. Suzanne works on the management of the construction industry in disaster environments. She has extensive research experience in disaster management, disaster recovery, and disaster reconstruction. Suzanne leads the Bush Fire Recovery project—a longitudinal study of the recovery and reconstruction following the 2009 Australian bush fires. She is also the founder and leader of the Centre for Disaster Resilience, Recovery and Reconstruction and a research leader with Resilient Organizations.

Chapter 1

Introduction: Urban Resilience Efforts in the Face of Natural Disasters

Pierre Filion, Mark Skidmore, and Gary Sands

The risk of suffering a disaster—a sudden catastrophic event that causes widespread destruction—is present in every city. Cities may be at varying degrees of exposure to a variety of natural disasters, including earthquakes, floods, tsunami and fires. While primarily the result of natural forces, such events are often exacerbated (and occasionally ameliorated) by the actions of urban residents. Unwise locational choices (flood plains, barrier islands or earthquake fault zones), inappropriate building construction techniques and high densities of development and population are choices people make that can contribute to the impacts of natural disasters.

As the global population continues to become increasingly urbanized (UN-Habitat, 2013), the potential for disaster losses increases as well. While the nature and frequency of many types of natural disasters has changed little over the past century, the extent of devastation that may be caused has escalated. The growth of urban populations has produced concentrations that are vulnerable to disasters. Urbanization has also augmented the extent and complexity of the built environment. When a disaster strikes an urban area, the potential for injury and loss of life is obviously a function of city size. There is more property subject to damage in cities than in thinly populated areas. A fire, flood or storm striking a rural area may damage structures and crops but restoration of these functions is relatively straightforward. The complexities and dependencies of the modern city make recovery of functionality more difficult.

The book is about urban resilience in the early twenty-first century. The term resilience has come into common usage in recent years (100 Resilient Cities, 2015; Pickett and Cardenasso, 2014; Vale and Campanella, 2005). Resiliency has been approached from a variety of perspectives, including urban design (Brown and Dixon, 2014), sociology (Holt, 2014) and climate change (Blakely and Carbonell, 2012; Newman and Beatley, 2009). Increasingly, resilience has become linked to sustainability (Coyle and Duaney, 2011; Roberts, Pearson and Newton, 2014).

In ecology, resilience refers to the capacity of an ecosystem to return to its original condition following a traumatic event. In current usage, the term has a much broader meaning. The 100 Resilient Cities organization, for example, defines resilience as the capacity of individuals, communities, institutions, businesses, and systems within a city to survive, adapt, and grow no matter what kinds of chronic stresses and acute shocks they experience (100 Resilient Cities, 2015). Thus, resilience often has a normative component, stressing not just returning to a prior state, but also implementing improvements.

The book draws its empirical substance from descriptions of recent natural disasters that have confronted urban areas. Disasters considered in the book comprise earthquakes, a tsunami, hurricanes, and floods which have taken place over the last decades. In one instance, however, the 1755 Lisbon earthquake, the book describes a historical event. The Lisbon case is introduced as a prototype for the coordinated state relief and reconstruction interventions in the face of severe natural hazards that have since taken place, including recent responses. The book casts light on variations in the reactions to the different disasters it discusses. It considers levels of preparedness, the mitigation of the consequences of hazards, the effectiveness of recovery instruments, and the attention given during the rebuilding phase to the avoidance of future destruction due to natural disasters. It also assesses the adequacy of the strategies deployed to address the disasters. Are these strategies suited to the nature and scope of the disasters? Have they minimized the effects of disasters and provided means to assure the safety and wellbeing of residents, while providing conditions for the rebuilding of more resilient urban forms?

Most chapters present a common scenario. The consequences of a disaster are often worse than they might have been otherwise due to insufficient preparedness. Inadequate preparation and response cause serious personal suffering and damage to the economy, which could both have been significantly reduced. And reconstruction does not always take place in a manner that will reduce damages in the eventuality of a future event of a similar nature to the one responsible for the destruction being repaired. The book does, however, introduce cases that have been more successful in alleviating the consequences of disasters. These cases demonstrate the benefits of preparedness as well as of the deployment of coordinated and well-funded recovery strategies. In one instance, effective efforts of this nature have involved primarily community-based self-help.

Preparedness for a natural disaster and the capacity to recover rapidly from the damages and destruction it inflicts are of course limited by the unpredictability of such events, the extent and complexity of required responses, and the availability of resources. Accordingly, any survey of disasters and the responses they have elicited will necessarily describe, along with some examples of successful preparedness and recovery, situations where preparation was wanting and aid to the population and subsequent recovery hampered by insufficient resources and/or inadequate decision making and coordination. After all, if readiness for and reactions to a natural event were perfectly honed, and, as a result, suffering and damages kept to a minimum, the trigger events would not qualify as disasters. They would rather be perceived as potential disasters that have been averted. But at the same time, full preparedness for all natural disaster eventualities would be far too expensive, and thus absorb too many resources that could be used in other areas of social life and, therefore, represent an unacceptable burden on society. It is a matter of balance. What measures in terms of preparation for natural events, the delivery of aid to victims, and overall recovery offer the best return for a given investment? What earthquake resistant construction techniques, land use planning and built forms that minimize flooding damages, along with evacuation strategies and efficient means of delivering aid can be adopted at an acceptable cost?

The book contributes to a growing conversation on how to deal with the destructive impact of natural events on cities. Why this increasing interest in natural disasters and their urban consequences? One obvious explanation is that, as the population of the planet expands and a growing proportion of people live in cities, more of the surface of the earth is occupied by urban development. On its own, this situation raises the odds that cities will be struck by natural events, and, in consequence, that large numbers of people will be victimized (Miletti, 1999; Pelling, 2003: 19–45). Moreover, the urbanization process in itself can exacerbate the implications of extreme natural events. Along with the severe impacts on densely populated areas of damages to buildings and infrastructures as well as resulting electricity, water and food supply interruptions, there are flooding risks accruing from the predominance of impermeable surfaces within cities (Brody et al., 2008). The presence in cities of land uses that are vulnerable to natural events and whose destruction endangers surrounding areas (and far beyond in the case of toxic spills and radioactive leaks) represents another urban vulnerability to natural disasters. This is the case notably of oil refineries and storage facilities, chemical industries, and, worst of all, nuclear power plants.

Increased interest for the effects of natural disasters on cities can also be attributed to a generalized perception that natural hazards are becoming more frequent and intense. If geological phenomena are for the most part outside the range of influence of human activity—although connections have been made between oil extraction, especially fracking, and earthquakes—the situation is somewhat different for climatic events such as hurricanes, storms, droughts, and periods of extreme temperature (Miller, Muir-Wood and Boissonnade, 2008; Stephenson, 2008). The increase in climatic disturbances scientists believe is caused by the atmospheric accumulation of greenhouse gases requires an adaptation of cities (Newman, Beatley and Boyer, 2009; Prasad et al., 2009). It indeed introduces an additional factor of unpredictability, which in turn increases interest in further study to reduce this uncertainty. For example, models of the past, where risks were assessed on the basis of the frequency of extreme events, are not as reliable as they used to be. Storms of an intensity that would be repeated every 50 or 100 years may now occur every 10 or 20 years, or adopt a random occurrence pattern (O'Brien et al., 2006). Given the evidence indicating that human activity is contributing to the frequency and intensity of climatic hazards, and the limited success of attempts at modifying behavior in this regard, it is justified to expect a further worsening of threats to cities from climatic events.

Yet another reason for heightened interest in the urban distress natural events cause is awareness of the threat such events can represent for technological systems supporting urban life. The failure of technological systems due to natural hazards can magnify considerably the gravity of these hazards. They can result in disruptions that endanger human life. For example, the January 1998 ice storm that affected southwest Québec and eastern Ontario provoked extensive and long-lasting power interruptions, forcing many people to seek refuge from the cold. At one point electric power came close to being lost in the Montréal water treatment plant, which would have left approximately 1.5 million people without drinkable water and, for many, the possibility of boiling water at home (Chang et al., 2007). There is also the risk of widespread and long-lasting contamination when nuclear power plants, oil refineries and chemical industries are damaged. Natural events, with impacts that would have otherwise been confined, become major sources of peril when they interface with technological systems (Boin and McConnell, 2007).

Like most research on disasters, this volume is mostly about resilience. How do city residents best to survive disasters? How do cities to bounce back and return to a prior state of system equilibrium? And, perhaps most importantly, how can city leaders and other policy-makers anticipate future disasters and take appropriate measures capable of alleviating their impacts? (Allenby and Fink, 2005; Haimes, 2009; Klein, Nicholls and Thomalla, 2003; Vugrin et al., 2010). To be effective, resilience strategies targeting natural hazards must be adapted to the threats confronting a given urban area. The measures suitable to a city located in an active earthquake zone differ from those appropriate to an urban area on a hurricane-prone coast (Kreimer, Arnolo, and Carlin, 2003). Still, the same fundamental steps will be found in all resilience strategies targeted at natural hazards: preparedness, response, and recovery (Gilbert, 2010). In the case of climate change, resilience could also include mitigation, in the form of limits on emissions responsible for the greenhouse effect and resulting climate extremes. While collectively such measures could eventually have a global impact, their effect on any given specific area will not be felt for a long time.

The process that has been relied upon to improve civil aviation safety could provide a model for addressing urban resilience failure in the face of natural hazards. The risk of flying accidents has been considerably abated through the carrying out of a formal investigation whenever an incident occurs, the quick dissemination of its findings, and the requirement that airlines conform to the ensuing recommendations from transportation safety agencies. The transposition of this model to the resilience of cities in the face of natural hazards could make it possible to identify causes of resilience failure as well as solutions to these problems. The dissemination of this knowledge and its integration to resilience planning would result in a reduction of the adverse impacts of natural hazards. There is a major obstacle to the application of this model to the urban resilience sphere, however: The model defines problems and solutions in purely technical terms, whereas decisions related to urban resilience are taken in a political economy context. In reality, even in the field of civil aviation, directives to improve aeronautical safety are not motivated exclusively by technical considerations. They do indeed take into account implementation costs and their consequences on the profitability of airlines.

The political economy perspective is preoccupied with decision making concerning priorities in the allocation of resources within society, along with the consequences of these decisions on economic development and wellbeing. At the heart of this perspective is the adoption of priorities to guide government responses to demands requiring the allocation of resources. Throughout history, societies have had to make trade-offs in the allotment of resources between immediate and long-term needs—eating grain now or keeping it for next year's sowing, for example. Also, with special relevance to preparedness for natural hazards, there is the issue of whether resources should be targeted exclusively at high probability risks or also be made available for less likely events. Market economies encounter endless debates regarding the amount of money that can be drawn from the private economy to be used by the state to provide collective goods such as education, healthcare, infrastructure and defense. In a market economy, governments must refrain from extracting too many resources from the private economy for fear that this will slow development and, consequently, have ill effects on the economy as a whole, and ultimately on the long-term capacity for the state to raise money (e.g., Drazen, 2000; Tabb, 1999; Usher, 2003).

Since the late 1970s, the persistent rise of neoliberalism has curtailed the importance of the state. This represents a challenge to reliance on the precautionary principle in policy-making, inherent in the concept of resilience. Neoliberalism promotes self-reliance, risk taking and values market processes over state intervention. It calls for the shrinkage of the state, as resources are primarily channeled towards market processes, especially the financial sector, which can itself be a major source of economic volatility as it occupies a larger share of the economy. What is more, neoliberalism entails a shift away from policy objectives that are not consistent with direct support of the market economy (Burgess et al., 1997; Duménil and Lévy, 2011). The risk-prone nature of neoliberalism, the accentuated position of financial capital within the economy, and a depletion of resources at the disposal of the state all contradict the precautionary principle and the achievement of urban reliance (Peck, 2010). There is an inherent contradiction between neoliberal values and their political and economic ramifications, on the one hand, and urban resilience, on the other (Filion, 2014).

It is important to realize that neoliberalism is one among different factors responsible for risk-prone behavior, lack of preparedness and insufficient recovery efforts. Another such factor is at the opposite end of the intervention spectrum from the neoliberal retrenchment of the state. It manifests itself when governments show a great deal of readiness in providing funding to repair damages inflicted by repeated disasters. The moral hazard resulting from such unrestricted compensation prevents a full consideration, when reconstruction takes place, of the likelihood of a repetition of prior natural disasters, as the financial risk associated with further such disasters is not borne by those who make redevelopment decisions.

It is unlikely that even under advancing neoliberalism, governments will totally neglect the fostering of resilience. There are two reasons for this. In the first place, for all its distrust of the state, neoliberalism values its role in securing the preconditions for an efficient operation of economic markets. The prevention of damaging disturbances to the economy caused by natural hazards constitutes an obvious precondition in this regard. In the second place, there are dire political consequences to the incapacity to prepare for, and respond effectively to, a natural disaster. Such a failure can reflect negatively on the public perception of an administration's capacity to govern (Stehr, 2006). The botched federal response to Hurricane Katrina became a blemish on the Bush administration, a mistake President Obama was careful not to repeat when Hurricane Sandy hit the northeast coast just before the 2012 election. Historically, natural hazards have been linked to changes of regimes and even the collapse of civilizations (Diamond, 2005). For example, abnormally cold weather conditions associated with El Niño and an Icelandic volcanic eruption, resulting in crop failures, are seen as contributors to the French Revolution (Grove, 1998; 2007; Wood, 1992). Given these political repercussions of disasters, one should not be surprised that in certain circumstances, a political calculus would lead to an overreaction to a disaster event, for fear that doing less would result in an electoral backlash (Stehr, 2006).

There is a powerful social inequity dimension to urban resilience. Individuals who have access to resources are more likely to reside in buildings that can resist natural events, can choose to live in urban areas that are less exposed to such events, and to possess the means to evacuate an area affected by a disaster. The experience of Hurricane Katrina has painted a vivid picture of how closely vulnerability correlates with poverty. Not only is access to private resources a factor of inequality in the face of natural disasters, but so are preparedness and recovery strategies. They too can reflect the uneven political and economic influence of different social groups, and the resulting lopsided treatment given by governments to these groups and the respective geographical areas they occupy (Berke, Kartez, and Wenger, 2008; Tierney, 2006).

Up to this point, attention has mostly centered on the efforts of the state at attempting to foster urban resilience. We now look at the role of other actors. Especially when the capacities of governments are overwhelmed by a disaster whose scale prevents them from engaging in recovery efforts that are commensurate with needs, other nations and international aid agencies can come to the rescue as was in the case for the 2004 Indian Ocean Tsunami and the 2010 Haiti earthquake. At the opposite end of the spectrum from supra-governmental involvement, are community-based self-help initiatives. These are especially required when effective government preventive and recovery measures are absent, which forces residents to rely on their own means to assure their survival and well-being in the face of natural disasters. Even when aid from governments and other agencies is available, community-based efforts are often better positioned than large organizations to provide a rapid response, such as immediate help to people caught in the rubble of an earthquake. In addition, residents are better aware of the needs of their local areas, such as the location of old people who could have problem withstanding a heat wave. As self-help community efforts tend to be a natural extension of existing social networks, mobilizing this kind of help can be problematic in the urban areas of economically and technologically advanced societies where such networks are often frayed (Putman, 2000).

While they are a serious source of damage and human suffering, natural disasters are not the only, or the most dangerous, risk confronting cities. A jet airliner falling from the sky into a populated area because of terrorist acts, mechanical failure or pilot error would be considered a human disaster. While disasters are typically associated with significant physical damage, disease and famine may in some cases be natural disasters that produce little damage to the built environment.

Human caused disasters, such as war and economic collapse, are indeed generally more devastating. Because, contrary to the case of wars and economic decline, damages inflicted by natural disasters happen over a brief time, from seconds to a few days, and because they do not stem from societal crises causing social disintegration, natural disasters are easier to address than these other forms of disaster. War and especially de-industrialization and capital flight can deprive a city of the conditions needed for its reconstruction. They also inflict more psychological distress than natural disasters (Norris et al., 2002). In fact, natural disasters accompanied by successful relief and reconstruction efforts can contribute to a feeling of ontological security. It is reassuring to know that in the event of a natural disaster effective relief is available and that damages to urban areas will be repaired. But such thinking about natural disasters does not account for the likelihood in the future of more frequent and extreme climate events triggered by greenhouse gases. Much of the forecast economic costs associated with climate change relate to urban destruction, disruptions and human injury resulting from hurricanes, storms, floods and extreme temperatures. The consequences of climate change will be felt worldwide, but with particular acuity in the most vulnerable countries, those especially exposed to changing climatic patterns and ocean rise, and without resources to respond adequately to the resulting crises. Accordingly, not only does the book offer critical perspectives on measures taken to address current urban disasters caused by natural events, but it also offers lessons that will be of increasing relevance as climate related hazards gain in intensity and frequency.

The chapters in the book are organized according to the stage of a natural disaster recovery process that they most emphasize. While a majority of chapters allude to more than one stage of such a response, they each give special attention to a specific stage. The sequence of the chapters is defined by to the attention they give to:

1. Preparedness (Chapter 2);
2. Immediate response to a disaster (Chapter 3);
3. Early recovery efforts to assure the survival of victims (Chapter 4);
4. Long-term strategies for the reconstruction of cities (chapters 5 and 6);
5. Efforts at bringing people back and re-launching the economy (chapters 7 and 8);

6. The evaluation of the costs, and thus of the economic ramifications, of a disaster (Chapter 9).

The first case focuses on community based rather than government led responses; the remaining cases, however, bring the critical role of government to the fore. Because of the scale of natural disasters, responses by local governments are seldom adequate by themselves; support from state and national governments, and in some instances from international organizations, is essential when disaster events occur. As several of the cases illustrate, governments at all levels have been ineffective when faced with large scale natural disasters. At each stage—from preventive planning through emergency relief and restoration efforts to long term reconstruction—we see the consequences of inadequate planning, poor communication and coordination, reliance on out of date or inadequate mitigation protocols.

Chapter 2, by Ersing, Alhassan, Ayivor, and Caruson, describes the reaction of two Ghanaian communities to a repeated natural hazard: recurrent floods. The first community demonstrates a strong mobilization around efforts at minimizing the impact of floods. These involve a regular clearing of drainage channels, the improvement of the water resistance of homes, and support from mutual aid networks in the face of danger. In contrast, the second community presents far less resilience capacity. The study attributes these differences to the joint effect of the intensity of social ties, education on community resilience, and leadership in the first community, and the absence, in the second community, of these conditions for locally based self-help. The chapter underscores the importance of community-based efforts when an area is confronted to a disaster, but also demonstrates that such efforts cannot be improvised, as they require the existence of a favorable social terrain from which to emerge.

While this chapter highlights the possibility that substantial differences can arise within the same contextual framework, it is important to recognize that external factors limit and shape the range of responses that are possible. The settlements described in this chapter have only limited access to financial aid from the national government or from insurance. The options available to these communities might have been considerably different if these or other external resources had been available.

The next chapter, by Schneider and Jordan, compares responses to two natural disasters: to the 2011 Fukushima nuclear meltdown caused by an earthquake-tsunami combination and the 2012 Hurricane Sandy. While earthquakes and tsunami are common in Japan, the nuclear incident that resulted constitutes the focusing event that sets this disaster apart from the others. It highlights the inadequacy of the response to the Fukushima nuclear incident, by pointing to a lack of preparedness and coordination. It blames political tensions within the Japanese government, insufficient communication between levels of government, and a cozy relationship between the nuclear power industry and the government agency overseeing them. This case provides a graphic illustration of how modern technology can multiply the effects of a natural hazard. Things were different in the case of Hurricane Sandy. The federal government located abundant emergency personnel in the target area before the hurricane struck and was quick to make financial assistance available for recovery and reconstruction. State and local agencies were then expected to address more local and immediate effects of the hurricane. While the chapter paints a predominantly positive picture of the response to Hurricane Sandy, it does point to gaps in the response of local administrations, some of which were overwhelmed by the scope of the disaster.

Chapter 4 by Pooyan concentrates on recovery efforts in the aftermath of the Kobe, Japan and Bam City, Iran earthquakes of 1995 and 2003. The attention is on emergency shelters and living conditions therein. It identifies a tension between the tendency on the part of aid agencies to locate the shelters away from destroyed urban sectors, so as to prevent interference with rebuilding efforts, and the preference on the part of people made homeless by the earthquake to remain close to their former residence. They indeed seek to maintain their social networks and keep an eye on their property. The chapter raises serious concerns that if reconstruction efforts are insufficient or too slow, temporary shelter areas may become permanent, which would become a long-term source of poor public health conditions and economic isolation for their residents.

The chapter by Feio, Bologna, Monteiro and Félix contrasts reconstruction strategies deployed in the aftermath of two earthquakes that took place in Portugal. By adopting as one of its case studies the massive 1755 Lisbon earthquake, the chapter deviates from the attention the book gives to recent natural disasters. Surprisingly, the 1755 case is presented as a best practice in terms of urban reconstruction, whereas the more recent one, the 1980 earthquake of Angra do Heroísmo in the Azores, is described as much less successful. In the view of this chapter, the Lisbon reconstruction strategy can be seen as the prototype of effective state intervention after a natural disaster. The Lisbon recovery and reconstruction clearly benefited from the concentration of power found in its monarchy, access to abundant resources, and effective leadership of Pombal. As a result, it took place rapidly and in a fashion that minimized suffering and improved the resilience of the urban environment. In contrast, in part because of its isolation, Angra do Heroísmo found it difficult to access the labor force and materials needed for its reconstruction.

Chapter 6, by Stevenson, Giovinazzi, Seville, Brown, Chang, and Wilkinson, discusses the Christchurch reconstruction choices made in the wake of the 2010 New Zealand earthquake. There is much in common between the Christchurch rebuilding decision making and the historical Lisbon experience described in the previous chapter. Both instances are characterized by a large-scale involvement of the state, the adoption of a highly coordinated strategy, and a commitment to rebuild the city in a fashion that will minimize damage from a future earthquake (see also Johnson and Olshansky, 2013). Chapter 6 also describes obstacles to the state-based reconstruction strategy. There is, for example, the disagreement of insurance companies with some aspects of the strategy. Another difficulty is the need to balance conditions for the immediate survival and wellbeing of people affected by the earthquake with long-term reconstruction objectives.

In Chapter 7, Kammerbauer addresses recovery efforts after Hurricane Katrina focusing on New Orleans's Ninth Ward. The chapter describes clashes between two contradictory recovery strategies: one promoting shrinkage and the other, growth. With the strategies come very different implementation measures, buyouts in one case and infill development in the other. The chapter blames the mixed message sent by these opposite strategies and the incompatibility of the measures they respectively put forth for the slow return of residents to the Ninth Ward. Another obstacle to recovery has been delays in providing funding for reconstruction. Two lessons emerge from the Ninth Ward case. The first is the need for closely coordinated recovery strategies. The second is the importance of a quick availability of reconstruction funds.

Chapter 8, by Kim and Marcouiller, discusses post-disaster resilience in the context of tourism. It emphasizes the importance of rapidly restoring confidence in a touristic area after it has been afflicted by a natural disaster. Otherwise, the area will suffer economically and find it difficult to recover from the destruction wrought by the disaster. The issue confronting such areas is that tourism is a discretionary expenditure and can thus easily be cancelled when circumstances seem to be unfavorable. Likewise, tourists can typically choose among many destinations, which makes it difficult for a city, which has been afflicted by a natural disaster and/or does not seem equipped to confront such an event, to compete for tourism with other, apparently more secure, sectors. Many tourist areas confront such circumstances, as they are often located in hot coastal regions, which are prone to hurricanes. Such tourism areas are therefore especially vulnerable to climatic events, which can have a devastating effect on their economic base, especially if these areas appear incapable of securing the safety of visitors and recovering quickly.

The chapter by Steindel discusses methods of calculating the costs of a disaster. Its approach introduces the reader to the far-reaching consequences of a disaster. The chapter compares different methodologies and shows how they yield very different evaluations of the costs of a same disaster. Chapter 9 focuses on the impact of Hurricane Sandy on New Jersey and pays particular attention to the effects of the hurricane on economic output. It illustrates the difficulties involved in accurately measuring the extent of damages and, hence, the possibilities of underestimating the true costs, which may lead to underinvestment in preventative measures.

References

100 Resilient Cities (2015). http://www.100resilientcities.org/#/-_Yz42MDo2MCdpPTEocz5j/.

Allenby, B. and J. Fink (2005). Toward inherently secure and resilient cities. *Science* 309(5737): 1034–1036.

Berke, P., J. Kartez, and D. Wenger (2008). Recovery after disaster: Achieving sustainable development, mitigation and equity. *Disasters* 17: 93–109.

Blakely, E. and A. Carbonall (2011). *Resilient Coastal Cities and Regions*. Cambridge, MA: Lincoln Institute of Land Policy.

Boin, A. and A. McConnell (2007). Preparing for critical infrastructure breakdowns: The limits of crisis management and the need for resilience. *Journal of Contingencies and Crisis Management* 15: 50–59.

Brody, S., S Zahran, W. Highfield, et al. (2008). Identifying the impact of the built environment on flood damage in Texas. *Disasters* 32: 1–18.

Brown, L.J. and D. Dixon (2014). *Urban Design for an Urban Century: Shaping More Livable, Equitable, and Resilient Cities*. New York: John Wiley & Sons.

Burgess, R., M. Carmona, and T. Koistee (1997). *The Challenge of Sustainable Cities: Neoliberalism and Urban Strategies in Developing Countries*. London: Zed Books

Chang, S.E., T.L. McDaniels, J. Mikawoz, and K. Peterson (2007). Infrastructure failure interdependencies in extreme events: Power outing consequences in the 1998 ice storm. *Natural Hazards* 41: 337–358.

Coyle, S. and A. Duaney (2011). *Sustainable and Resilient Cities: A Comprehensive Action Plan for Towns, Cities, and Region*. Hoboken, HJ: John Wiley & Sons.

Diamond, J. (2005). *Collapse: How Societies Choose to Fail or Survive*. New York: Penguin.

Drazen, A. (2000). *Political Economy in Macroeconomic*. Princeton, NJ: Princeton University Press.

Duménil, G. and D. Lévy (2011). *The Crisis of Neoliberalism*. Cambridge, MA: Harvard University Press.

Filion, P. (2014). Fading resilience? Creative destruction, neoliberalism and mounting risks. *Sapiens* 6(1): http://sapeins.revues.org/1523.

Gilbert, S.W. (2010). *Disaster Resilience: A Guide to the Literature*. Washington, DC: US Department of Commerce, National Institute of Standards and Technology (NIST Special Publication 1117).

Grove, R.H. (1998). Global impact of the 1789–1793 El Niño. *Nature* 393: 318–319.

Grove, R.H. (2007). The great El Niño of 1789–1793 and its global consequences: Reconstructing an extreme climate event in world environmental history. *The Medieval History Journal* 10: 75–98.

Haimes, Y.Y. (2009). On the definition of resilience in systems. *Risk Analysis* 29: 498–501.

Hoffman, L.M. (2014). From sustainability to resilience: Why locality matters, in W. Holt (ed.) *From Sustainable to Resilient Cities: Global Concerns and Urban Efforts*. Bingley, UK: Emerald Group, 341–157.

Holt, W. (ed.) (2014). *From Sustainable to Resilient Cities: Global Concerns and Urban Efforts*. Bingley UK: Emerald Group.

Johnson, L. and R. Olshansky (2013). The road to recovery: Governing post-disaster reconstruction. *Land Lines*, July 14–21.
Klein, R., R. Nicholls, and F. Thomalla (2003). Resilience to natural hazards: How useful is this concept. *Environmental Hazards* 5: 35–45.
Kreimer, A., M. Arnolo, and A. Carlin (eds) (2003). *Building Safer Cities: The Future of Disaster Risk*. Washington, DC: The World Bank.
Miletti, D. (1999). *Disasters by Design: A Reassessment of Natural Hazards in the United States*. Washington, DC: Joseph Henry Press.
Miller, S., A. Muir-Wood, and A. Boissonnade (2008). An exploration of trends in normalized weather-related catastrophe loss, in H. Diaz and R. Murnane (eds) *Climate Extremes and Society*. Cambridge: Cambridge University Press, 225–247.
Newman, P., T. Beatley, and H. Boyer (2009). *Resilient Cities: Responding to Peak Oil and Climate Change*. Washington, DC: Island Press.
Norris, F., M. Friedman, P. Watson, et al. (2002). 60,000 disaster victims speak: Part 1. An empirical review of the empirical literature, 1981–2001. *Psychiatry: Interpersonal and Biological Processes* 65: 207–239.
O'Brien, G., P. O'Keefe, J. Rose, and B. Wisner (2006). Climate change and disaster management. *Disasters* 30: 64–80.
Peck, J. (2010). *Constructions of Neoliberal Reasons*. Oxford: Oxford University Press.
Pelling, M. (2003). *The Vulnerability of Cities*. London: Earthscan.
Pickett, S.T., M.L. Cadenasso, and J.M. Grove (2004). Resilient cities: Meaning, models, and metaphor for integrating the ecological, socio-economic, and planning realms. *Landscape and Urban Planning* 69: 369–384.
Prasad, N., F. Ranghieri, F. Shah, et al. (2009). *Climate Resilient Cities: A Primer on Reducing Vulnerabilities to Disasters*. Washington, DC: The World Bank.
Putman, R. (2000). *Bowling Alone: The Collapse and Revival of American Community*. New York: Simon and Shuster.
Roberts, P., L. Pearson, and P. Newton (2014). *Resilient Sustainable Cities: A Future.* Abingdon, UK: Routledge.
Stehr, S. (2006). The political economy of urban disaster assistance. *Urban Affairs Review* 41: 492–500.
Stephenson, D. (2008). Definition, diagnosis, and origin of extreme weather and climate events, in H. Diaz and R. Murnane (eds) *Climate Extremes and Society*. Cambridge: Cambridge University Press, 11–23.
Tabb, W.K. (1999). *Reconstructing Political Economy: The Great Divide in Economic Thought*. London: Routledge.
Tierney, K. (2006). Social inequality, hazards, and disasters, in R. Daniels, D. Kettl, and H. Kunreuther (eds) *On Risk and Disaster: Lessons from Hurricane Katrina*. Philadelphia, PA: University of Pennsylvania Press, 109–128.
UN-Habitat (2013). *Time to Think Urban*. Nairobi: United Nations Human Settlement Programme.
Usher, D. (2003). *Political Economy*. Hoboken, NJ: Wiley.

Vale, L. and T. Campanella (2005). *The Resilient City: How Modern Cities Recover from Disasters*. New York: Oxford University Press.

Vugrin, E.D., D.E. Warren, M.A. Ehlen, and R.C. Camphouse (2010). A framework for assessing the resilience of infrastructure and economic systems, in K. Gopalakrishnan and S. Peeta (eds) *Sustainable and Resilient Critical Infrastructure Systems*. Berlin: Springer-Verlag, 77–116.

Wood, C.A. (1992). The climate effect of the 1783 Laki eruption, in C.R. Harrington (ed.) *The Year without a Summer?* Ottawa: Canadian Museum of Nature, 58–77.

Chapter 2

Enhancing Hazard Resilience among Impoverished Urban Communities in Ghana: The Role of Women as Catalysts for Improvement

Robin L. Ersing, Osman Alhassan, Jesse Sey Ayivor, and Kiki Caruson

Introduction

Women play an integral role in the ability of a community to mitigate, prepare for, and recover from a disaster due to their central function as primary care providers for both the young and the aging (Poole, 2005). In many countries women not only fulfill the role of caretaker to these vulnerable populations, but their own vulnerability increases dramatically in the context of disaster events. In the aftermath of a disaster event, women survivors often encounter harsh and sometimes violent conditions such as increased exposure to domestic and sexual violence, lack of access to relief services, inadequate compensation for losses, sustained economic hardship, long-term displacement, lack of appropriate health care, and a lack of voice in recovery and rebuilding efforts. These outcomes are exacerbated in impoverished communities.

An important emergency management concept is that all disasters are local; their greatest impact is on a specific location that exists within a broader social, economic and political context. As local events occur, when women fail to recover from disaster, entire communities suffer as well. Women emerge as catalysts for improved disaster management when they are able to actively participate in preparedness, mitigation, response, and recovery roles. One simple yet powerful way that women contribute to the resiliency of their communities is by leveraging their social ties to inform, educate, and mobilize individuals within their personal networks and across their communities.

This chapter analyzes two urban communities in the African nation of Ghana. The Sabon Zongo is an urban settlement situated on the western edge of the metropolis of Accra, the capital city of Ghana; Yeji is an urbanized fishing settlement located in the Volta Basin Region of Ghana, 455 kilometers north of Accra. Both communities are deeply impoverished and lack basic services important to the health and safety of residents.

The field research conducted for this analysis represents a collaborative project between the University of South Florida and the University of Ghana to engage in the development of a cross-national, community-based disaster resilience initiative to empower and equip women as local-level leaders in disaster risk reduction. The Global Hazard Resilience through Opportunities for Women (GHROW) initiative is designed to build capacity around disaster risk reduction knowledge, skills, and advocacy so women can use their social and communication networks to inform, educate, and mobilize others within their communities.

Literature Review

Globally, individual communities are faced with varying degrees of risk imposed by the impact of hazards. Hazard threats are the product of the intersection of community-specific factors (vulnerability) and the type of hazard (Tobin and Montz, 1997). Disaster vulnerability research examines geophysical processes as well as social, cultural, political, and economic factors to better predict the level of risk for a particular community (Cutter and Finch, 2008; Tobin and Montz, 1997; Wisner et al., 2004). Wisner et al. define vulnerability as "the characteristics of a person or group and their situation that influence their capacity to anticipate, cope with, resist, and recover from the impact of a natural hazard" (2004, 11). Disasters are the result of poorly managed interactions between hazards and human society that have exposed people to harm without adequate coping mechanisms. In many cases, society has exacerbated the impact of such hazards through development activities (Hewitt, 1997; Wisner et al., 2004). As noted in the Hyogo Framework for Action 2005–2015, disaster-prone developing countries, especially least developed countries and Small Island Developing States (SIDS), warrant particular attention in light of their higher vulnerabilities and risk levels; these levels often exceed their capacities to mitigate, prepare for, respond to, and recover from disasters. The need for building capacities to increase a region's resilience to mitigate hazard outcomes has become pressing.

Resilience can be defined as a high level of preparedness or readiness to contain or endure a disaster event, as well as the ability to recover or pull back from the shocks and destruction of the disaster event. Disaster resiliency, thus, involves the process of building these preparedness and recovery processes in local communities so they become institutionalized, scaled up, and tailored to specific and general disaster events (ADPC, 2006). Disaster resiliency includes the roles of social connections, familial ties, and community networks in supporting people, processes, and places when confronted with exposure to a hazard.

Global Urbanization and Disaster Risk

There is currently an explosion of global urbanization, with an anticipated five billion people expected to reside in urban communities by the year 2030 (United Nations Population Fund [UNPF], 2007). The African continent leads this unprecedented urban growth. Such rapid population shifts in a geographic location give cause to examine the potential for increased levels of disaster risk. Environmental health factors such as sanitation, general hygiene, potable water, adequate shelter, and transportation are just a few conditions that can adversely affect the ability to mitigate the impacts of a hazard (UNPF, 2007). Although increased urbanization contributes to concentrated poverty and its related social and environmental problems, urban localities may also provide opportunities to build individual and collective capacity, thereby promoting the development of policies to address educational and economic conditions (UNPF, 2007). Urban centers are able to provide sources of both formal employment, through industrialized jobs, as well as outlets for informal employment. Oftentimes, however, the informal economy of urban settlement areas falls short of being able to provide a sustainable livelihood, particularly for women (International Labour Office, 2009).

The connections between urbanization, disasters, and vulnerability in Africa are staggering. Indeed, the annual urban growth rate in Sub-Saharan Africa is almost 5 percent, twice as high as in Latin America and Asia (UNDESA, 2012). Sub-Saharan Africa also has the world's largest proportion of urban residents living in slums; 72 percent of Africa's urban citizens reside in slums, representing an estimated 187 million people. As more and more people seek a better life in towns and cities, partly as a result of climate change in the solely rain-fed agricultural sector, the slum population in urban Africa is projected to double every 15 years, illustrating the "urbanization of poverty" process (UNDESA, 2012). African cities in the new millennium are thus confronted with the problem of accommodating rapidly growing urban populations in inclusive cities, providing them with adequate shelter and basic urban services, and ensuring environmental sustainability as well as enhancing economic growth and development (UNISDR, 2013).

Urbanization and Disaster Risk in Ghana

Since 1970, the shift from rural to urbanized living in the African nation of Ghana has been a steady process, with significant urban population growth beginning in 1990 and continuing over the last two decades (COHRE, 2008). As a result, the population in the capital city of Accra is expected to reach 6 million people by 2015. As noted, increased urbanization has been positively correlated with a heightened disaster risk. Natural and human induced hazards negatively impact infrastructure and assets and increase risk, especially among already vulnerable groups such as migrants, women and children, the aged, and persons with disabilities.

In Ghana, as in many places in Africa, the urban poor often live and work in places with high exposure to environmental hazards due to the lack of basic sanitation services. Confronted to high population densities and climate change effects, the poor are the most vulnerable to environmental disasters and shocks.

One major effect of urbanization in Ghana is the growing incidence of slum development. A contributing factor has been a limited regulatory framework, thought to be indifferent or hostile to the needs of the poor. In 2001, the number of people living in slums in Ghanaian cities was estimated at 5 million and growing at a rate of 1.8 percent annually. The metropolitan areas of Accra, Kumasi, Sekondi-Takoradi, Tema, and Tamale are locations with pronounced slum development in Ghana (National Development Planning Commission, 2005). In most cases, migration into urban areas exceeds the availability of affordable housing services. As a consequence, squatters illegally occupy available land (both public and private), thereby undermining regulatory mechanisms and reducing the economic value of land and its potential use. Such squatter settlements are often perceived as hot spots of poverty, crime, and disease. In addition, rapid urbanization in many Ghanaian communities has contributed to increased congestion and chronic traffic problems, slow mobility, illegal tapping of water and electricity lines, and dwindling health care resources (Mutiara et al., 2008).

Women, Urbanization, and Disaster Risk

Women and their children are among the poorest in urban locales, yet these areas can provide greater opportunity for females to develop diverse social ties and connections to obtain access to services that would otherwise be unobtainable (UNPF, 2007). Although the literature on the impact of global urbanization on the social, economic, and health conditions of women continues to emerge, several factors are significant. Roughly half of all global migrants are women, many of whom relocate to urban areas without a husband or family (COHRE, 2008). Increasingly, females have taken on the responsibility as key wage earners, thus providing the primary economic means for their families. Given their limited resources and work skills, a large portion of women find shelter in urban slums where opportunities of an informal economy offer potential income; albeit, these earnings are rarely enough to lift an individual out of poverty. Despite their considerable drawbacks, urbanized settlements are considered by many women as a path out of rural poverty and an early step toward gender equity and empowerment. In many nations, urban locales offer greater social mobility among women, as their collective interactions allow an exchange of resources and knowledge and the hope that they might contribute to local decision making (UN-Habitat, 2006).

Global urbanization poses a unique set of challenges and opportunities for women; so too do natural and man-made hazards, particularly when they impact urban locales. The inequitable access to, and distribution of, resources needed to prepare for and recover from a disaster creates a disproportionate risk for females. Global disasters such as floods, earthquakes, droughts, tsunamis, and hurricanes paint a bleak picture for the welfare of women. Reports from several international organizations have identified a disproportionately more difficult path to recovery for women in the aftermath of a disaster, including a higher death rate (UNISDR, 2010, World Bank, 2012).

In Ghana, natural disasters not only worsen economic deprivation in the short-term, but also compromise the long-term welfare of women by forcing affected households to use or sell assets that would otherwise be used to meet basic needs and contingencies (Caruson et al., 2014). A lack of basic sanitation in many urban communities is a significant cause of disease after events such as floods—a common occurrence in urban settlements. Arguably, many people who die from cholera and other water-borne diseases could have survived if local and national governments ensured that residents had access to potable water, sanitation, and adequate health services (Alhassan, 2014).

The Urbanized Settlements of Sabon Zongo and Yeji

The two urban sites selected for in-depth study include: Sabon Zongo, an unplanned settlement located in the western part of metropolitan Accra, the capital city of Ghana, and the Volta Lakeshore settlement, located in Yeji in the Brongo Ahafo Region of Ghana.

Sabon Zongo—Accra

Sabon Zongo is a low lying area with an estimated terrain elevation of around 114 meters above sea level. With an estimated population of 18,616 in 2000, the area recorded 1,252 house compounds with 4,156 households. The population grew to an estimated 70,000 persons by 2012. The settlement of Sabon Zongo is home to an immigrant community of primarily Muslim residents (from Mali, Burkina Faso, and Ghana's northern regions) totaling 25,000. Sabon Zongo is one of 12 communities within Greater Metropolitan Accra that rank lowest in terms of environmental health and disaster monitoring. It is characterized as "absolutely deprived" of basic services such as potable water, waste disposal, sanitation, drainage, pest control, and pollution abatement. Coupled with a crowded physical infrastructure including houses, shops, roads and markets, sanitation and waste management are a serious challenge.

Sabon Zongo is a richly diverse community; dominantly Muslim, it consists of "foreign" groups such as the Hausa, Fulani, Yoruba, Zabrama, Wangara, Mossi, Kotokoli, Busanga, Gao, and Kanuri, as well as Ghanaian groups such as the Dagomba, Wala, and Gonja. Non-Muslim indigenous groups such as the Akan, Ga, and Ewe also reside within the settlement. Despite the variety of languages spoken daily in Sabon Zongo, no fewer than 15, residents often enjoy strong social, religious, and historical connections with the community. Though other areas of Accra have received considerably more government investment, many generations have chosen to remain in the Sabon Zongo (Pellow, 1991; Nicola Mousset-Jones, 1999).

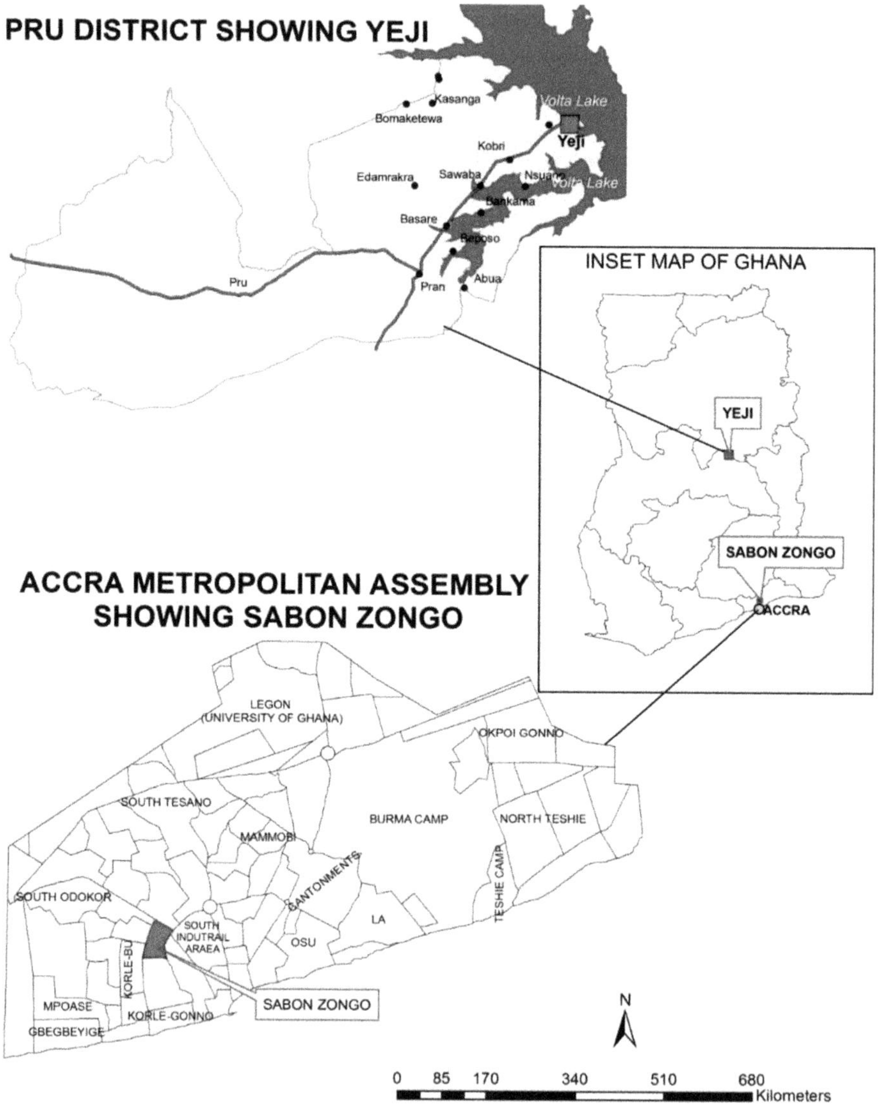

Figure 2.1 Location of Sabon Zongo and Yeji urban settlements
Source: Authors.

Each ethnic group within Sabon Zongo has its own chief; however, the power of the head chief follows the line of the Bako family, of the Hausa group. Residents appreciate being able to live in peace because there is a local native court to settle community disputes through arbitration (Pellow, 1991). The main public facilities are mosques; Sabon Zongo had approximately 40 as of 2013.

There is one basic public school, shops, many descendant-owned compounds, a fuel filling station, and a movie theatre called Gaskia. Women and men participate in several traditional organizations including zumunci and zumanta, which are Islam-based yet ethnically mixed and provide aid to the Sabon Zongo population (Pellow, 1991). Traditional occupations of men in the settlement include trade in cows, sheep, goats, or kola nuts. A substantial number of men find work as drivers, mechanics, and casual workers in the Makola and Agbogbloshie markets. There are few social and community workers in health, education and other social protection services. Women engage in home-based micro-enterprises including petty trading, food processing, and food and water vending. Girls can be found selling assorted wares including iced-water and traditional medicines within the settlement.

The settlement of Sabon Zongo suffers from all the problems associated with urban vulnerability, including infrastructure and housing stressors, economic deprivation, serious environmental and health burdens, and inequality in the distribution of public amenities (Songsore et al., 2009). An exception is the organizational skills women in the community often display by mobilizing women's groups to undertake household and community projects which result in substantial improvement in conditions (sanitation, roadway access) within the settlement.

Yeji-Volta Lake Basin

Yeji is an urbanized lake port located on the eastern shore of Volta Lake, situated along the Accra-Atebubu-Salaga corridor, linking the northern parts of Ghana to the South. It is the Pru District capital within the Brong-Ahafo Region of Ghana. The strategic location of the city and its elevation to a district capital in 2004 have made it an important commercial hub, as well as an attractive destination for migrants from other parts of Ghana. As a result, Yeji experienced a surge in population from 18,593 in 2000 to more than 35,000 by 2013. During the past decade, flooding has been a major problem throughout the Volta River Basin in general, in Yeji in particular. The problem has been attributed to several factors: rainfall variability, the creation of uncoordinated dams upstream in neighboring Burkina Faso without appropriate management practices, inappropriate land-use practices, and overcrowding.

Fishing, farming, and associated businesses dominate the economy of the area. As the district capital, there are also public and private sector employers including local government, decentralized agencies, utility services, educational institutions, medical facilities, and security services. The informal employment sector includes petty traders, artisans, and private transport operators.

Yeji has 12 "electoral areas" (the smallest political unit within Ghana's electoral system), each with representation at the District Assembly. The lakeshore settlement communities of FanteAkura No. 1 and FanteAkura No. 2 are located within the Jinebinbisa Electoral Area. These two communities are situated in designated flood-prone zones deemed unsafe by the Volta Basin Authority (VRA). Volta Lake was created as a result of the impoundment of the Volta River at Akosombo for hydroelectric power generation in 1961.

The VRA passed a law prohibiting human activities and settlement establishment below the 280 foot contour line and within a 50 meter buffer zone. Due to a lack of monitoring and enforcement of the law, and the absence of visible landmarks such as boundary pillars defining the area of the buffer zone, large scale encroachment has taken place. Migrants have settled within the buffer zone in FanteAkura Nos 1 and 2, under the guise of unawareness.

Residents in FanteAkura Nos 1 and 2 live in poorly constructed mud houses which provide little protection against floods. The living conditions are similar to those found in Sabon Zongo, and include: lack of basic sanitation systems and potable water, unmarked and clogged roadways, disease, and poverty. The residents consider the lakeshore area a commercial enclave where they are able to eke out a livelihood despite the vulnerability of the area to natural disasters. For example, devastating floods were recorded by Ghana's national disaster agency, NADMO, in the Volta Basin in 2007 and 2010. During the floods of 2010 official records from the district NADMO office indicated that 360 houses were flooded and over 1,000 residents displaced in FanteAkura Nos 1 & 2. Despite the devastation caused by floods and stern warnings from government officials not to settle along the shore line, new structures continue to be erected in the area.

Methods

Focus groups are especially relevant in Ghana, a country with a strong oral tradition. In addition, the terrain and geographical composition of the communities involved make in-person data collection the best method for assessment. Focus group discussions allow for robust conversation and for the identification of issues that are local and specific to individuals, as well as those that transcend individual and group perspectives.

A series of focus groups in the two designated urban areas were interviewed in November of 2013. In the Sabon Zongo settlement, two separate focus group interviews were conducted and facilitated by the Janok Foundation: a group of 15 women identified as homemakers and a second group of 11 identified as women engaged in small business or petty trading. In the Yeji settlements, two separate focus group interviews were conducted: the first included 23 participants

(FanteAkura No. 1) and the second included 16 (FanteAkura No. 2). All community participants had experienced an extreme event—flooding and/or fire.

Focus group questions addressed four areas of evaluation: (1) the strengths or assets of women in the context of disaster mitigation and preparedness; (2) the strengths or assets of women during the recovery phase of disaster management; (3) the specific needs or vulnerabilities of women in the context of disaster events; and (4) how women utilize their social ties and networks to strengthen their resiliency to disasters.

An identical set of questions were posed to all focus group participants, and focus group conversations were conducted and recorded in the local language of the region and translated to English. In addition to the reactions of the women to discussion questions, information was collected regarding each woman's age, level of education, marital status, occupation, and ethnicity. Overall, 65 women across two urban communities contributed their opinions to assessments of disaster resiliency among women in highly impoverished communities. The focus group participants ranged in age from 20 years to over 60. Their education levels varied widely across each group, as did their marital statuses, although the majority of the women reported being married. Several of the focus groups were exclusively or largely comprised of Muslim women. Ethnicities ranged from Hausa, Fulani, Zabarima and Ewe, to Gonja, Adidome, and Akan.

Results: Themes across the Sabon Zongo and Yeji Settlements

Within the two urban sites selected for in-depth study, Sabon Zongo and the Yeji lakeshore communities of FanteAkura No. 1 and FanteAkura No. 2, floods, heavy rainstorms, and fires (house fires associated with indoor, open flame cooking and/or electrical fires) are the most commonly identified hazards. In addition, the settlement communities are impacted by environmental health risks such as a lack of potable water, sanitation systems, and hygiene facilities. These communities have generally been excluded from any meaningful planning in terms of the built environment, environmental conditions, and health services. Even where services are provided, a lack of financial and human capacity at the metropolitan government level limits service provision. In these settlement communities, it is women and children who suffer the most from inadequate or unavailable services and amenities.

Disaster Preparedness Capacities

The women who participated in the focus group conversations identified various capabilities and strengths women have to aid in disaster preparation. In Sabon Zongo, the most widespread mitigation activity is clearing drainage channels to enhance the passage of water and to reduce flooding effects on homes and other structures. In Sabon Zongo, Ms. Jane Oku, a former local assembly member and the founder of the Janok Foundation, has emerged as a powerful leader among women and men in the context of disaster resiliency. Oku is responsible for a number of non-governmental (NGO)-funded improvements to the community, including dedicated drainage channels, public toilets, and garbage containment. Keeping the tangle of roadways free of debris is a critical mitigation activity.

Encroachment consistently challenges proper drainage within the community; even a brief rain storm can cause flooding in areas of the settlement where drainage is compromised. In Sabon Zongo, women regularly participate in clean-up exercises to clear the drainage channels of filth and debris. Clean-up exercises just prior to heavy rains normally attract more women and other community members. Recently, working tools such as spades, rakes, wheel barrows, and long brooms were provided by the sanitation company Zoomlion Company, Ltd. for drainage improvement activities.

The women of Sabon Zongo have an able champion in Jane Oku. Smart and tenacious, Ms Oku demonstrates the impact that a single individual can have on a community. She is an inspiration to the women of the settlement and many of them draw their strength from her example. She has transferred skills and knowledge to many women in Sabon Zongo, who then share this information with the wider community. Oku's efforts are inclusive of men as well.

Within a deeply impoverished community, she has generated awareness and acceptance of the importance of simple mitigation activities—collecting refuse in cans or designated areas instead of littering the streets with all manner of items, for instance. Despite its urban location, cattle herders are relatively common in Sabon Zongo. Cow dung is often disposed of in drainage channels, which prevents drainage and causes flooding during rains. Indiscriminate disposal of human waste is also a common practice which focus group participants considered a serious health threat, especially during periods of flooding. In Yeji, a lack of facilities for the proper disposal of human waste was also identified as a serious sanitation hazard—exacerbated during periods of flooding.

Disaster preparedness for rainstorms and flooding in the Yeji settlements is centered on protecting physical structures. Most women mentioned the need to anchor building roofs with nylon ropes to concrete blocks on the ground. This simple technique works to keep roofs from blowing away during storms. Discarded fishing nets are used over thatch roofing to prevent the thatch cover from being blown away. Some families have also hardened their homes by creating ridges along the outsides of structures to slow the force of flood waters; others have started to use sandcrete (a mixture of sand and concrete) instead of mud and small branches to construct dwellings.

Absent a charismatic and adept female leader such as Jane Oku, the women of Yeji glean strength from the need to protect their lives and livelihoods. They identify themselves as mothers who have a responsibility to protect their children and as petty traders who must protect their financial position. A two-fold investment was articulated in this context; short-term investment in petty trading which provides an economic livelihood day-to-day, and a long term investment in their children's education as a means of escaping poverty.

Disaster Recovery Capacities

Among the women of the Sabon Zongo and Yeji settlements, resourcefulness is the key to survival. Compared to their counterparts in Yeji, the women of Sabon Zongo had greater access to outside sources such as government agencies and NGOs for assistance with recovery efforts. When asked about their most pressing recovery needs, the women from Sabon Zongo discussed immediate or short-term needs, whereas the women from the Yeji settlements identified longer term goals. In the wake of a disaster, the women of Sabon Zongo voiced a desire for emergency medical care, food, and drinking water. These services and items are immediate needs in the context of disaster events. The women also professed an interest in curbing the outbreak of disease. Lack of a medical clinic in the community was a major source of worry to many Sabon Zongo focus group participants. Disease and malnutrition complicate the recovery effort. The women of Yeji spoke about housing-related needs, including access to cement and roofing sheets and the availability of micro financing to support home improvements. Given the extent to which floods destroy the mud houses that provide shelter for the vast majority of the residents in the Yeji settlements, their focus on the longer term process of hardening structures is not surprising.

The women representing Yeji voiced a desire to relocate further from the lakeshore, but all lacked the requisite resources to do so. Adequate—even temporary—shelter for children and property is a primary concern. The scarcity of micro loans for the purchase of hardware and materials to strengthen older structures and to build newer structures that are better able to resist disaster events is viewed as a major hindrance to both preparation and recovery efforts by women in Yeji. Recovery from disasters in Yeji is dependent on a woman's ability to contract loans from individuals or from banks. For those who cannot access loans, they continue to live in unsafe structures after the disaster and concentrate their energies on business activities to mobilize resources. Recovery in both communities is a long and difficult process.

Leveraging Social and Familial Networks to Improve Disaster Resiliency

Women in Sabon Zongo have close social ties, a reflection of the tight-knit nature of the larger settlement. Women have organized several women's associations or groups, primarily for the pooling of resources for events such as weddings and births, but funds have also been collected to make small loans or to distribute money to families in severe need. The focus group participants strongly agreed on the importance of their social and familial networks as a mechanism for assistance.

The women perceive these support systems as far more important to their ability to withstand disasters than the activities of external organizations. In the event of an emergency, neighbors generally alert each other through loud shouts. Upon hearing the shouting, neighbors and community members routinely offer help. During floods, women assist each other to salvage personal belongings from flooded homes, work collectively to clear drainage channels, and sweep flood waters from homes and other structures.

It was clear during focus group discussions that the social networks women in Sabon Zongo have developed are critical to their preparedness and to recovery efforts. In a few instances, disaster victims have benefited from the benevolence of wealthier individuals within the community, but overall, their efforts at improving disaster resiliency are focused on leveraging social ties within the community more effectively. These initiatives include enfranchising more women into social groups/organizations; enlarging the scope of association activities; more organized and collective mitigation activities; and establishing formal outlets for knowledge sharing and communication. Information and knowledge regarding disaster experience is passed from person to person through word of mouth. Through neighbor-to-neighbor relationships, community members share successful and unsuccessful strategies.

For instance, women from Sabon Zongo learned from each other that soapy water can be used to effectively put out fires. This is a simple, locally relevant strategy for a settlement where the condition of the roads and the lack of street signs make it virtually impossible for municipal fire trucks to provide assistance.

The Yeji settlements lack the cohesive spirit evident in Sabon Zongo. According to the women who participated in the focus groups, during emergency events all the members of the community are affected, thus creating a situation where it is difficult to help each other. Unlike the situation in Sabon Zongo where neighbor-to-neighbor relationships and networks among women serve as a source of strength, apart from the ad hoc funeral support groups which exist in Yeji, there are no social or women's groups to assist during times of disaster. Family ties are important in Yeji and many of the women spoke of receiving help or donations from relatives, but their support networks rarely extended beyond the scope of their family members. Instead of looking to one another for assistance, the women frequently spoke of the need to borrow money from external organizations such as banks. Community members do share ideas, knowledge, and skills, but the outlets for such discussions are more limited than in Sabon Zongo. In comparison to Sabon Zongo, in the Yeji settlements the ability to utilize social networks for disaster resiliency is, at this time, limited.

Male Community Members

In terms of the involvement of men in general, and husbands specifically, in disaster management support networks, women in both communities offered different evaluations depending on the culture/religion of the participant. In both Sabon Zongo and Yeji, preparedness and recovery activities fall largely to the female population; however, within Sabon Zongo, there is a coordinated effort to include men in disaster management endeavors. The women's groups have a slogan: "Men, Rise Up in Support of Women!" that they often chant as they go about their work within the community. Their strategic effort to enfranchise men is the product of the reality of living in a highly conservative and patriarchal society. While women manage the bulk of mitigation and preparedness activities, during periods of recovery from disaster events (flooding and fires), men in Sabon Zongo often help to clear choked gutters and remove debris while women generally sweep and clean. In some Sabon Zongo homes the cooperation and assistance from men is noteworthy.

Similarly, in Yeji, opinions differ regarding the contribution of men during disaster events. According to the women interviewed, some men are very helpful, whereas others are not. In some families, women discuss preparedness activities before flooding events with their husbands. In other homes, women alone bore the responsibility. The Yeji settlements can claim a male leader who works to mobilize young people and women for self-help initiatives. A former member of the district assembly, Mr Yahaya Ibrahim, is a community activist who has led several initiatives to improve the safety of living conditions of residents living along the lakeshore. His efforts are not focused solely on women, but as a group, women benefit from his advocacy.

Discussion and Conclusion

In both settlement locations, Sabon Zongo and FanteAkura Nos 1 and 2 in Yeji, flooding is a consistent hazard and one which causes dislocation, disease, and in some cases death. Several focus group participants have experienced the death of a family member as a result of flooding. As key stakeholders in disaster preparedness and recovery, women are critical to disaster resiliency efforts. Their opinions, knowledge, and skills, however, are too often an underutilized asset. Activism among women in Sabon Zongo demonstrates the potential of social networks to strengthen disaster management efforts.

The focus group conversations have provided insight into the development of a community-based, disaster resilience "toolkit" used to empower and equip women as leaders in disaster risk reduction through active participation in preparedness, response, and recovery roles. The Global Hazard Resilience through Opportunities for Women (GHROW) initiative promotes best practices and capacity building strategies to support women as positive agents of change in disaster risk reduction. The GHROW toolkit is designed to advance knowledge, skills, and advocacy so women can use their social ties to inform, educate, and mobilize others within their social networks. The goal is to cultivate an environment where women's support systems are commensurate with their level of disaster responsibilities. Local relevancy is a fundamental component of the evolution of the toolkit.

Given the information collected from focus group participants, three areas for capacity building are critical: (1) skills related to data collection, information analysis, reporting and documentation of activities, and record keeping; (2) methods for encouraging and promoting the further development of existing social networks, with the goal of creating forums where networks across communities are able to collaborate with each other in the exchange of ideas and knowledge; and (3) strategies for enhancing awareness and understanding of modest but innovative methods for improving and sustaining community resiliency.

By analyzing the opinions and perceptions of women in two highly vulnerable urban settings in Ghana, this chapter provides a better understanding of the capacities of women in these communities (as well as many others) to enhance disaster resiliency. Women are increasingly adopting innovative strategies, both as individuals and as a collective, to prepare in advance for an emergency rather than mobilizing after a disaster strikes. The evidence from Sabon Zongo highlights the ability of women to organize and leverage their social networks to improve disaster resiliency.

The exchange of disaster knowledge and information among women cultivates a record of best practices which can be institutionalized in their communities, as well as replicated in other localities with similar vulnerabilities. Scaling-up women's community networks is critical, especially in places such as Yeji where such organization is nascent. Through personal, social, and community ties, women contribute to expanding core competencies, capacity building, and coordination efforts for the benefit of the entire community in the context of disaster resiliency.

References

Alhassan, O. (2014). Environment and development in Sub-Saharan Africa, in Takyiwaa Manuh and Esi-Sutherland-Addy (eds) *Africa in Contemporary Perspective: An Undergraduate Textbook.* Accra: Sub-Saharan Publishers, 134–154.

Asian Disaster Preparedness Center (ADPC) (2006). *Critical Guidelines: Community-Based Disaster Risk Management*, www.adpc.net.

Caruson, K., O. Alhassan, J.S. Ayivor, and R. Ersing (2014). Disaster and development in Ghana: Improving disaster resiliency at the local level, in N. Kapucu and K.T. Liou (eds) *Disaster and Development: Examining Global Issues and Cases*. Berlin: Springer, 273-287.

Center on Housing Rights and Evictions (COHRE) (n.d.). *Women, Slums and Urbanization: Examining the Causes and Consequences*. Geneva, Switzerland: COHRE.

Cutter, S.L. and C. Finch. (2008). Temporal and spatial changes in social vulnerability to natural hazards. *Proceedings of the National Academy of Sciences* 105: 2301–2306.

Dacosta, A., T. Dari, and J. Koomson (2013). Risk perception and disaster management in the Savannah Region of Ghana. *International Journal of Humanities and Social Science* 3(3): 1–12.

Government of Ghana/National Development Planning Commission (2010). *Medium Term National Development Policy Framework: Ghana Shared Growth and Development Agenda (GSGDA), 2010–2013*. Accra: National Development Planning Commission.

Hesseling, G. (1994). Children as social security in Africa? *Poverty, and Development*. The Hague: Ministry of Foreign Affairs, 129–136.

Hewitt, K. (1997). *Regions of Risk: Hazards, Vulnerability and Disasters*. London: Longman/Pearson.

International Labour Office (2009). *The Informal Economy in Africa: Promoting Transition to Formality Challenges and Strategies*. Geneva, Switzerland: International Labour Organization.

McEntire, D.A. (2004). Development, disasters and vulnerability: A discussion of divergent theories and the need for their integration. *Disaster Prevention and Management* 13(3): 193–198.

Mousset-Jones, N. (1999). "Mapping a microcredit programme for women of the SabonZongo." Best Paper Series, Symposium on Global Issues in Public Administration. Syracuse NY: Maxwell School of Citizenship and Public Affairs.

National Development Planning Commission (2008). "A National strategy for sustainable human settlements in Ghana." (Draft Report) Accra: Government of Ghana.

Pellow, D. (1991). The power of space in the evolution of an Accra Zongo. *Ethnohistory* 38: 414–450.

Poole, M. (2005). *Family: Changing Families, Changing Times*. Crows Nest, Australia: Allen and Unwin.

Songsore, J., J.S. Nabila, Y. Yangyuoru, et al. (2009). Integrated disaster risk and environmental health monitoring; Greater Accra Metropolitan Area, Ghana, in M. Pelling and B. Wisner (eds) *Disaster Risk Reduction: Cases from Urban Africa*. London: Earthscan, 65–85.

Tobin, G.A. and B.E. Montz (1997). *Natural Hazards: Explanation and Integration*. New York: The Guilford Press.

UN-Habitat (2006). *State of the World's Cities: 2006–2007*. Nairobi, Kenya: United Nations Human Settlements Programme.

United Nations Department of Economic and Social Affairs (UNDESA) (2012). *World Population Prospects, the 2010 Revision*, http//esa.un.org/unpd/wpp/index.htm.

United Nations International Strategy for Disaster Reduction (UNISDR) (2010). *Disaster Risk Reduction: An Instrument for Achieving the Millennium Development Goals.* New York: United Nations.

United Nations International Strategy for Disaster Reduction (UNISDR) (2013). *Global Assessment Report on Disaster Risk Reduction.* Geneva: UNISDR.

United Nations Population Fund (UNPF) (2007). *State of World Population 2007: Unleashing the Potential of Urban Growth.* New York: UNPF.

Wahab, E.O. and U.C. Isiugo-Abanihe (2008). An assessment of importance of children in old age security provisions among the Ijebus of South-West, Nigeria. *African Journal for the Psychological Study of Social Issues* 11: 46–63.

Wisner, B.P., P. Blaikie, T. Cannon, and I. Davies (2004). *At Risk: Natural Hazards, People's Vulnerability and Disasters.* 2nd edition. London: Routledge.

World Bank (2010). *Report on the Status of Disaster Risk Reduction (DRR) in Sub-Saharan Africa.* Washington, DC: The World Bank.

World Bank (2012). *Making Women's Voices Count: Integrating Gender Issues in Disaster Risk Management.* Washington, DC: The World Bank.

World Bank (2014). *World Development Report—Risk and Opportunity; Managing Risk for Development.* Washington, DC: The World Bank Group.

Chapter 3
An Analysis of Governmental Performance during Urban Disasters: Fukushima and Hurricane Sandy

Saundra K. Schneider and Marty P. Jordan

On March 11, 2011, a magnitude 9.0 earthquake struck off the north-eastern coast of Japan. A 13-meter-high tsunami followed, inundating 560 km^2 along Japan's coast, leveling infrastructure, and paralyzing the communities in its path. The Great East Earthquake-Tsunami killed nearly 19,000 people and left over 150,000 homeless. The earthquake-tsunami also affected the Fukushima Daiichi nuclear power plant, cutting off external power and incapacitating the emergency generators designed to cool the reactors during a crisis. Consequently, three of the reactors' cores overheated, triggering hydrogen explosions and causing massive radiation pollution. The government created an exclusion zone around the plant to protect local residents, yet winds carrying radioactive particles reached as far as Tokyo, Japan's capital city and largest urban center with a population of 30 million people. Tokyo is two hours south of the Fukushima Plant. Radioactivity levels increased 30 times above normal and almost forced a citywide evacuation. Experts estimate that it will take more than 40 years to decontaminate the Fukushima Daiichi site. The earthquake-tsunami nuclear meltdown is responsible for an unprecedented series of problems in close proximity to the one of the world's most populous metropolitan areas.

About a year and a half later, a monster storm named Hurricane Sandy headed for the East Coast of the United States. The storm made landfall in southern New Jersey on the evening of October 29, 2012. Hurricane Sandy battered communities in New York and New Jersey with heavy rains, strong winds, and record storm surges. Then, it continued to wreak havoc over a 1,000-mile radius across a dozen states along the eastern US coastline. Hurricane Sandy's impact was extraordinarily large and severe. It caused extensive flooding to homes and businesses, disrupted communication channels and transportation systems, and left massive amounts of debris almost everywhere in its path. Thousands of people sought shelter in temporary facilities and over 9 million residents lost power. The storm was responsible for more than 150 deaths. Hurricane Sandy's impact was magnified because it struck densely populated, intensely developed regions along the eastern US seaboard, a region where much of the built environment conformed to older codes and standards such as the National Flood Insurance Program and local ordinances enacted to limit flood damage.

Both of these events—the earthquake-tsunami-nuclear meltdown in Japan and Hurricane Sandy along the north-eastern coast of the United States—affected a sizable number of communities across large geographic areas. They also created unprecedented problems for the urban populations and metropolitan centers in their wakes. Public and private organizations from local to national levels were called upon to deal with the disruptions caused by these disasters. The scope and severity of these events strained the capabilities of governmental response systems in both countries. Although the Japanese and American governments have allocated tremendous amounts of money to facilitate recovery efforts, some affected communities still have not received the assistance necessary to rebuild. Disaster victims continue to express frustration about how public officials have handled the situations. As a result, critical questions have been raised about these governments' ability to address major disasters, particularly when they affect large urban areas.

This chapter examines the effectiveness of governmental disaster recovery actions subsequent to the earthquake-tsunami-nuclear meltdown in Japan and Hurricane Sandy along the north-eastern corridor of the United States. We focus on these two events for several reasons. First, both Japan and the United States routinely face a wide variety of disasters (hurricanes, earthquakes, floods, terrorism, oil spills), and they have established fairly elaborate intergovernmental mechanisms for dealing with such situations. Second, the two disaster situations affected large urban populations, raising serious concerns about the ability of governmental response systems in both countries to handle such disasters in a quick and effective manner. Finally, these two situations provide a number of important insights into the determinants of effective government performance during crises and the consequences of these actions for the successful recovery of cities and urban populations when disaster strikes. As such, they demonstrate the vulnerability of large urban areas to major, catastrophic events, as well as the need to develop more effective strategies for handling these situations.

This chapter is organized as follows. First, we recount the March 2011 earthquake-tsunami disaster and ensuing Fukushima Daiichi nuclear meltdown calamity in Japan, its effect on urban populations and cities, and the governmental reactions at the national, prefecture, and city levels. We highlight the political, economic, social, cultural, and geographic factors affecting the response. (See also Perrow.) We also offer some lessons learned from the successes and failures of the Fukushima recovery efforts. Then, we follow the same trajectory for Hurricane Sandy's impact along the East Coast of the US in October 2012. Although the events surrounding Hurricane Sandy are different from those that occurred in Japan, they do reveal many of the same problems and issues. We conclude by offering recommendations in order to capitalize on the strengths of existing emergency response processes, as well as identifying ways to address the shortcomings of these approaches in the face of the next urban disaster.

The Fukushima Nuclear Crisis

A magnitude 9.0 earthquake—Japan's strongest on record—struck at 2:46 pm on March 11, 2011, 70 km off the east coast of the northern Oshiku Peninsula. Remarkably, considering the quake's magnitude, the submarine temblor was responsible for marginal loss of life and moderate damage. This is largely due to Japan's restrictive building codes requiring earthquake resistant infrastructure and emergency preparedness for seismic activity. However, the seaquake induced a 13-meter-high tsunami that crashed ashore on Japan's north-eastern coast 40 minutes after the extreme seismic activity. The wall of seawater flooded 560 km^2 along Japan's Tōhoku coastline leveling and damaging 1.2 million homes, buildings, businesses, and factories (Al-Badri and Berends, 2013; Vervaeck and Daniell, 2012); destroying roads and bridges; wiping out communication lines and transportation infrastructure; leaving millions of tons of debris; and inflicting over US $200 billion in property damages (Carpenter, 2012). Even more tragic, the Great East Japan Earthquake-Tsunami claimed the lives of nearly 19,000 people, left thousands injured, and more than 150,000 homeless (Carpenter, 2012). Those living in coastal cities and towns were particularly paralyzed by the domino-disaster. The elderly, poor, and vulnerable populations were the most affected by the disaster, unable to evacuate to higher ground in time (Shrader-Frechette, 2012). In fact, 65 percent of the victims who lost their lives were over the age of 60 (Mimura et al., 2011).

The earthquake-tsunami also affected the Fukushima Daiichi No. 1 Nuclear Power Plant (Fukushima Plant), located 45 km south of Sendai City, the largest city in the region with 1 million people. The plant is 250 km north of Tokyo, Japan's capital and metropolitan area, with a population of 30 million people. The earthquake severed connections with the electrical grid, cutting off external power to the plant, while the tsunami flooded operations, incapacitating 12 of the 13 emergency generators designed to cool the plant's six reactors during crises. Reactors 1, 2, and 3 were in operation at the time of the quake, while reactors 4, 5, and 6 were out of operation for maintenance and upgrade purposes, yet still highly vulnerable. Without external and backup power to run the reactors' cooling system, pressure and temperatures in the first four reactors began to rise (Carpenter, 2012; Eisler, 2013; Funabashi and Kitazawa, 2012).

Workers from the Tokyo Electric Power Company (TEPCO), the plant's operator, responded to the crisis by pumping fresh water into the reactors' chambers. When that proved ineffective, and with a drastically deteriorating situation in Reactor 1, the decision was made to open pressure vents. While the pressure did drop, a hydrogen leak mixed with oxygen and caused an explosion that destroyed Reactor 1's containment building. Both the venting and explosion released massive plumes of radioactive material into the environment. Attempting to stabilize the situation, plant workers proceeded to further vent pressure buildup and injected millions of gallons of available fresh water, and then seawater, to cool the reactors' cores. However, rising temperatures, increasing pressure, and lowering water levels in the reactors precipitated a second explosion in Reactor 3 on March 14 and then a third explosion in Reactor 2 on March 15 (which led to a fire in Reactor 4).

Nuclear meltdowns occurred in the plant's first three reactors. The meltdowns, explosions, pressure venting, exposed fuel rods, and spent fuel pools all contributed to dangerously high levels of radiation around the plant and in the local area (Carpenter, 2012; Ohnishi, 2012). As a result, 230,000 residents living within 30 km of the plant were forced to evacuate. If the meltdowns had produced more damage to the reactors' containment vessels, the release of radioactive materials would have been so great that residents living within 250 km of the Fukushima Plant would have needed to evacuate (Funabashi and Kitazawa, 2012). This zone included urban residents in Tokyo. Such an evacuation would have been unprecedented. Although in the end Tokyo did not need to be evacuated, strong winds carried radioactive particles as far as Tokyo. Radioactivity levels in the capital reached 30 times the normal level (Carpenter, 2012).

Higher than normal levels of radioactivity affected the entire region's air, water, soil, and agricultural produce (milk, beef, fish, apples, rice, and vegetables). Products from the Fukushima prefecture were pulled from store shelves throughout the country. The local apple, fish, cattle, vegetable, and rice industries suffered great financial losses, topping one billion dollars (Carpenter, 2012). In addition, storage complications and leaks in the containment vessels resulted in TEPCO releasing hundreds of millions of gallons of radioactive water back into the ocean. Radioactivity levels in ocean samples were thousands of times greater than the legal limit. All of this raised concerns about the enduring environmental and health effects for the region and its millions of residents. While the nuclear accident itself did not result in the loss of human life, plant employees suffered much higher than normal radiation exposure. Many workers fell ill due to radiation poisoning or were injured by reactor explosions (Carpenter, 2012). The long-term health consequences to plant employees and inhabitants living in nearby cities and towns is still unknown (Christodouleas et al., 2011).

In the first few weeks after the nuclear calamity, Japan's Nuclear and Industrial Safety Agency (NISA) officials put the severity level at 5 on the International Nuclear and Radiological Event Scale (INES). But radiation leaks, management missteps, hundreds of powerful aftershocks, and increased risks to densely populated areas continued to slow recovery efforts and heighten safety concerns. In turn, NISA revised the alert upward to its highest level, 7 (Al-Badri and Berends, 2013). Only the Chernobyl nuclear disaster had previously reached this level of severity. Due to the earthquake-tsunami and nuclear crisis, the national government shut down all of Japan's 54 nuclear facilities. This led to severe electricity shortages and extended blackout periods throughout the country (Carpenter, 2012). Moreover, the Japanese public lost confidence in the safety of nuclear power. In fact, following the disaster, nearly three-quarters of the public was in favor of eliminating or reducing reliance on nuclear energy, with most supporting a phase-out overtime (Shrader-Frechette, 2012).

It took TEPCO months to contain the accident. In 2014, TEPCO is still in the initial phase of trying to clean up the Fukushima Plant and compensate affected residents—more than three years after the incident occurred. Complete cleanup and demolition of the site is expected to take at least 45 years (Al-Badri and Berends, 2013). To be sure, the Great East Japan Earthquake-Tsunami was a rare event of extreme proportions, but it was not unprecedented. An earthquake-tsunami of this magnitude is estimated to hit Japan once every thousand years, with the last of this size occurring in 869 AD (Mimura et al., 2011). As a result, the ensuing disaster at the Fukushima Plant, although triggered by the earthquake and tsunami, could have been anticipated and averted.

The Japanese Governmental Disaster Response System

Japan is no stranger to natural and man-made disasters affecting urban populations. Japan's set of islands sits in the Pacific Ocean's ominous "Ring of Fire," where 90 percent of the world's earthquakes occur (USGS, 2012). Four tectonic plates converge on Japan, making it a hotbed for seismic activity. Japan accounts for 20 percent of the world's earthquakes with magnitudes greater than 6.0; it also carries a risk four times greater than other developed countries for landslides (Hood, 2012; Kingston, 2014). Moreover, Japan's mountainous regions cover 70 percent of its territory, forcing higher population density along the coastal low lands and cities, augmenting the risk and vulnerability of millions of residents living below sea level and in flood-prone areas (Mimura et al., 2011). This risk is further exacerbated by the fact that, on average, three typhoons strike Japan each year (Hood, 2012).

Several severe natural disasters affecting densely populated areas have marked Japan's history. In September 1923, an estimated 7.9 magnitude seabed earthquake hit 50 km south of Tokyo, claiming the lives of 140,000 people (Ishibashi, 2004). In January 1995, a 7.2 Richter Scale earthquake struck Kobe, Japan, a port-city of 1.5 million residents, killing 6,400 (Shaw and Goda, 2004). Torrential rains from a typhoon hit Japan's Tokai region in September 2000, triggering massive floods and landslides, displacing 58,000 persons (Ikeda, Sato, and Fukuzono, 2008). In fact, paleo-tsunami and historical seismology research has documented at least 70 devastating tsunamis and over 400 destructive earthquakes to hit Japan since 600 AD (Ishibashi, 2004; Nöggerath, Geller, and Gusiakov, 2011).

Although with less frequency than natural disasters, Japan has also experienced multiple man-made disasters. For example, in an act of domestic terrorism, a coordinated series of five sarin gas attacks hit the Tokyo subway system in March 1995. Thirteen were left dead and thousands injured (Okumura et al., 1998a).

These disasters—natural and man-made—brought to light many of the deficiencies in Japan's response readiness in urban areas. For example, the national government only dispatched 2,000 Japanese security defense forces (SDFs) to Kobe, a city of 1.5 million residents, to assist in relief efforts following the 1995 earthquake. During the sarin gas attacks in 1995, Tokyo's metropolitan first responders, emergency medical services, and hospitals were ill-trained to deal with a chemical exposure of this magnitude. Although the national government and SDFs had the tools and capability to deal with such a chemical crisis, they played a limited role in this scenario (Okumura et al., 1998a; 1998b; 1998c).

Nonetheless, these recent disasters served as "focusing events" (Birkland, 2007) for policy-makers to revisit, strengthen, and make structural changes to Japan's disaster response system at the local, regional, and national levels. Learning from these missteps and building on past success, Japan has developed one of the most sophisticated disaster response systems in the world (OECD, 2009). Most crises in Japan are handled at the city level, with local authorities as first responders with prefecture or federal government support if necessary. This "bottom-up" approach enables local authorities with direct knowledge of the situation to respond quickly and effectively. However, larger-scale disasters that span multiple jurisdictions or exceed the technical and logistical capacity of local authorities require a more centralized, "top-down" response (Schneider, 2005; 2011; Sylves, 2008). Depending on the gravity of the disaster, the national government, including the Prime Minister's Office and federal agencies (SDF, Fire and Disaster Management Agency, Police Agency), may take charge in order to coordinate efforts between national, prefecture, and city entities.

The Japanese disaster response system is also one of the most technologically advanced in the world. Japan implemented the System for the Prediction of Environmental Emergency Dose Information (SPEEDI) in 1984 to collect and disseminate predictions of radioactive exposure to aid in evacuations (Funabashi and Kitazawa, 2012). In 2002, the federal government completed a network of 700 high-sensitivity motion observation stations across inland and coastal Japan in an attempt to better detect seismic activity and predict future earthquakes (Obara et al., 2005). In 2006, Japan launched a participatory flood-risk communication system (Pafrics) to improve risk assessments as well as educate and coordinate disaster prevention and response between urban residents, regional authorities, and private organizations (Ikeda, Sato, and Fukuzono, 2008). And the Japanese government has invested in engineering and constructing coastal dykes and cement breakwaters to protect urban coastal communities from tsunamis and other weather-related events. Although these barriers did not stop the March 2011 tsunami, they did help lessen the loss of life and damages (Mimura et al., 2011).

The Response to the Fukushima Nuclear Crisis

Initially, city, regional, and national responses to the earthquake-tsunami disaster unfolded quite well. Following the earthquake, local officials from coastal cities issued evacuation alerts via loud speakers and radio transmissions telling residents to move to higher ground. Even before these alerts, administrators and teachers at elementary, middle, and high schools, prepared through drills, evacuated children to higher places. This saved the lives of nearly 3,000 students in Kamaishi alone

(Mimura et al., 2011). At the national level, before the tsunami hit, the Prime Minister's office had officially established an Emergency Disaster Response Headquarters and mobilized SDFs. More than 107,000 SDFs were on the ground within a week of the earthquake strike providing disaster assistance, a marked improvement from the response time of the 1995 Kobe earthquake (Ministry of Defense, 2012).

Nonetheless, despite Japan's advanced warning system and quick reaction to the earthquake-tsunami event, the Japanese government's handling of the nuclear crisis at the Fukushima Plant was nothing short of calamitous. Poor risk management coupled with miscommunication (and in some cases, deliberate deception) between TEPCO, the national government, and local authorities led to a series of major missteps. Simultaneously managing the response to the mammoth earthquake-tsunami disaster and the response to the nuclear crisis was challenging and unprecedented. Adding to the complication, the Japanese government tried to work in tandem with TEPCO, the largest private utility company in Japan, to resolve the situation. TEPCO often failed to share information, downplayed the severity of the situation, and even tried to cover-up mistakes. Furthermore, less than one week into the crisis, TEPCO proposed abandoning the plant, which would have put nearby reactors and millions of residents at grave risk. This proposal was quickly rejected by Prime Minister Kan.

The calamitous reactions to the crisis were also due to *structural problems* in the nation's emergency management system. The Japanese government and TEPCO were simply unprepared and ill-equipped to deal with the nuclear crisis and its impact on urban populations. Mistakes were made, one after another. For starters, when on-site plant workers turned to the TEPCO emergency manual for guidance in responding to the crisis, the out-of-date manual offered no suggestions for how to deal with extended blackout periods. Even worse, Japan's Nuclear Safety Commission (NSC), the entity responsible for nuclear safety, did not specify appropriate steps to deal with extended power outages as they were considered improbable events (Fukushima and Kitazawa, 2012).

These problems surfaced because of the private sector's close interpersonal relationship with government regulatory agencies in Japan. Much has been written about the Japanese "Iron Triangle," where the Liberal Democratic Party (LDP), the bureaucracy, and the business community worked in tandem to grow Japan's economy (Sakakibara, 2003). The LDP has dominated postwar politics, building a strong bureaucracy that advances rather than regulates the private sector. In fact, both NISA and NSC, the agencies charged with ensuring the safety of nuclear power, were embedded in the Ministry of Economy, Trade, and Industry (METI), which promotes nuclear energy use. Further contributing to poor regulation is the traditional practice of *amakudari*, where civil servants retire from the bureaucracy and take top management positions with private and public firms they once oversaw (Carpenter, 2012). Large utility companies like TEPCO that engaged in *amakudari* oftentimes got a slap on the wrist for major violations.[1] This regulatory environment allowed TEPCO to maintain seawalls that were too low and inadequate waterproof storage for backup generators at the Fukushima Plant (Lipscy and Kushida, 2012).[2]

Another factor that contributed to poor oversight and supervision was an over confidence in the safety of nuclear energy and the capabilities of the nation's technologically advanced disaster preparedness system. Utility companies convinced the Japanese public and politicians of nuclear power's "absolute safety" (Funabashi and Kitazawa, 2012). To be sure, many scientists warned of the dangers of nuclear energy and of possible cataclysmic events (Kingston 2014; Mimura et al. 2011). Even TEPCO had previously determined, through geological and paleo-tsunami studies, that the Tōhoku region was a high-risk area (Nöggerath, Geller, and Gusiakov, 2011). Despite these warnings and other international nuclear crises (Three Mile Island in the US, Chernobyl in Ukraine), many citizens believed that nuclear disasters would not happen in Japan, and if they did, they would be contained quickly.

Similarly, there is a strong societal belief that Japan's technological advancement and global status is sufficient to withstand any disaster. The integrated and technologically sophisticated disaster system appeared to be infallible. For example, prior to the Kobe earthquake in 1995, city residents believed they were safe from tremors because the region had never been hit by a quake before (Shaw and Goda, 2004). Prior to the sarin gas attacks, "the Japanese believed that the Tokyo subway system was the safest transportation system in the world" (Okumura et al., 1998: 613). Indeed, Japan's mitigation and preparedness strategy was largely based upon prior disaster experiences, and not the potential for a new, unforeseen disaster. This hubris contributed to an inability (or unwillingness) to consider "worst-case scenarios" ('t Hart, 2013).

Inexperience also hampered the response to this crisis. Given the political dominance of the LDP, the Democratic Party of Japan or the DPJ (the main opposition party) had little track record of handling major disasters (Krauss, 2013). Moreover, Prime Minister Kan and the DPJ campaigned for a reduction in the size of government, alienating the bureaucrats that would subsequently play a critical role in helping to resolve the situation. Prime Minister Kan, especially skeptical of

1 TEPCO had faced dozens of whistleblower complaints (even from the prefecture's governor) concerning the safety of the Fukushima Plant's operations and maintenance. While some of these complaints did result in the temporary suspension of operations at the plant, NISA oftentimes just passed the complaints on to TEPCO (Carpenter, 2012).

2 Lipscy and Kushida 2012 detail the variation in preparedness of nuclear power plants within Japan (i.e., not all plants were as vulnerable as the Fukushima Plant), comparing Japanese facilities' preparedness to other seaside nuclear plants around the world. They found many of the Japanese facilities' protection measures to be inadequately low given the historical risk of severe tsunamis. The plants particularly at risk were older and operated by Japan's largest utilities, including TEPCO.

the bureaucracy's ability to react quickly, named himself the key decision maker and often micromanaged aspects of the response. Kan's actions, however, were somewhat justified. A day into the crisis when Kan asked NSC Chairman, Haruki Madarame, if there was a risk of explosion at the plant, Madarame rejected any such possibility. Hours later when Reactor 1 exploded due to an oxygen leak, Kan lost confidence in Madarame. Nonetheless, Kan's approach caused unnecessary delays. He insisted on visiting the Fukushima Plant less than 24 hours into the crisis to meet with the plant's director and assess the situation himself. Kan's visit held up the venting of reactors, distracted plant personnel, and slowed response activities at a critical time (Al-Badri and Berends, 2013). Moreover, unconvinced by Madarame and others' initial reactions to the possibility of recriticality from seawater injections (whereby melted fuel debris could go back to criticality, that is, uncontrolled nuclear chain reaction), Kan waited several precious hours before authorizing the cooling of reactors with seawater. On this issue, the plant's director actually disobeyed Prime Minister Kan and TEPCO management's orders to halt seawater injections. While many have praised the plant director's decision to continue seawater injections (since it helped lessen the gravity of the situation), others have commented that unilateral decisions outside the chain of command could be confusing for personnel involved in the response (Shinoda 2013).

In addition, miscommunications, misinformation, and delays in information sharing often exacerbated the crisis. On multiple occasions, TEPCO management underreported certain levels of radiation, the amount of contaminated material, or the gravity of the crisis, only to significantly increase revised figures later. In addition, TEPCO failed to share necessary information with federal, prefecture, and city officials. For example, local police and fire fighters helped to inject water into the reactors, yet TEPCO refused to provide them a map of Fukushima Plant's facilities (Al-Badri and Berends, 2013). Reports of misinformation by TEPCO are still occurring now, three years after the disaster occurred.

Moreover, several hours passed before the decision was made to vent Reactors 1 and 2. Yet, even after the decision was made and even after Prime Minister Kan's on-site visit, plant personnel did not act for another seven hours (Funabashi and Kitazawa, 2012). This further increased the pressure and hydrogen levels in the containment vessels, causing the explosion in Reactor 1. Even evacuation and stay-indoors orders for residents were unclear and provided insufficient information as to where to go, creating confusion and chaos (Al-Badri and Berends, 2013). And the SPEEDI system, designed to detect and predict radiation levels to aid in evacuations, was not used until nearly a month into the crisis (Funabashi and Kitazawa 2012).

Politics also had a negative effect. Prime Minister Kan's position was already considered politically vulnerable, with a confidence vote scheduled before the earthquake-tsunami hit. The calamitous response to the nuclear crisis further exacerbated his vulnerability. The LDP rejected the Prime Minister's invitation to hold a cabinet seat and help in the disaster response and reconstruction. And the political animosity between the parties led to delays in getting aid to the affected region. Instead of working together, "each of the major actors involved in the crisis continue[d] to pursue, protect, and defend their own self-interests" (Krauss, 2013). The nuclear crisis revealed the deep fissures that exist in Japan's political system.

Urbanization also contributed to the fumbled emergency response. Since 70 percent of Japan is covered with mountains, most of the population lives in valleys and near the coasts. Although nuclear facilities are generally built in less-populated areas, the topography and high population density of Japan mean that nuclear power plants are never far from urban areas. Due to limited space, many plants' reactors are built too close to one another (Funabashi and Kitazawa, 2012). This was also the situation with the Fukushima Plant. The close proximity of the reactors contributed to the domino effect of explosions and meltdowns. All of this increased the vulnerability of the nation's urban populations, which created additional challenges for the public and private institutions involved.

Lessons Learned from the Response to Fukushima: Successes and Failures

According to an independent investigation by Japan's parliament, the Fukushima nuclear crisis "was a profoundly manmade disaster—that could and should have been foreseen and prevented ..." (BBC News, 2012). Both TEPCO and the Japanese government were found culpable in failing to mitigate and prepare for the disaster, as well as adequately respond to the crisis. The report also identified Japan's culture and political system as further culprits (BBC News, 2012). The cozy relationship between regulators and utility companies, which led to oversight of looming violations, along with a belief in the absolute safety of nuclear energy and the infallibility of Japan's disaster response system, contributed to the crisis. Indeed, substantial evidence suggests that while the earthquake and tsunami were unavoidable, the ensuing calamity of events at the Fukushima Plant were preventable.

There is no question that the scope and magnitude of the crisis required a centralized governmental response. Millions of residents and urban communities were (and continue to be) at great risk. Early on in the disaster, decision making was centered in the Prime Minister's Office in an effort to facilitate better coordination across and within the governmental system. But, during rapidly evolving disaster situations, like the Fukushima nuclear crisis, a very centralized response can be problematic. It can take too much time to channel all the information and requests to the top of the command chain ('t Hart, 2013). "Effective responses in such extreme circumstances are necessarily improvised, flexible, and networked (rather than planned, standardized, and centrally led). They are driven by the initiative of operational leaders and the strength of pre-existing ties between the teams and organizations they represent" ('t Hart, 2013: 7). Because of Prime Minster Kan's micromanaging, on-site expertise was lost and critical decision making delayed. This, in turn, further exacerbated (rather than mitigated) the crisis. A prior campaign to marginalize the bureaucracy, as well as miscommunication, misinformation, and inexperience, led to distrust between the individuals and agencies involved in the process.

The nuclear crisis also revealed that Japan's disaster response system remains highly reactive and fragmented between federal, prefecture, and city authorities (OECD, 2009). To ensure a successful response, capacity building, education, and training are needed. Local first responders need to be trained in dealing with a nuclear meltdown situation. And communication needs to be improved across all segments of the disaster response framework. Again, with so many governmental agencies involved, and intergovernmental cooperation required for major crisis events, operational leaders need to be adaptable, empower their response teams, and communicate across all entities involved.

Preparedness for a disaster, however, needs to extend beyond a disaster plan. Regulatory agencies should also be autonomous and separate from ministries that promote economic growth and energy use. Independence would allow regulators to press operators on violations (building reactors too close to one another, low seawalls, and unsafe working conditions) and work with utilities (and all private companies) to prepare for known and unforeseen situations. This also includes making sure that personnel are not simply rotated throughout agencies, frequently changing positions, and gaining little experience or training in emergency response. Institutional knowledge is essential for a successful response. Similarly, the practice of *amakudari*, or any tradition of a revolving door between the public and private sectors, does little to improve the capabilities of bureaucratic agencies involved in emergency relief operations.

Perhaps most importantly, more attention should be devoted to preparing for large-scale, multiple disaster scenarios. The Japanese public, policy-makers, civil servants, and businesspersons need to overcome the cultural and societal belief of Japan's disaster response infallibility. The Fukushima Plant crisis revealed that there are, indeed, serious problems with the Japanese approach to large-scale emergencies. Disasters can strike when and where they are least expected. This is why preparation, planning, and drills are critically important in facilitating successful crisis responses. Emergency response plans need to be flexible and updated on a continuous basis in order to meet the rapidly changing conditions that can develop during an emergency situation.

There must also be more consideration given to large-scale events occurring in heavily populated urban areas and communities. The Fukushima Plant crisis put tens of millions of Japanese residents at risk. It would be difficult enough coordinating a response if a nuclear meltdown occurred in a remote part of the world, but it is extremely difficult when such an event happens in a densely populated area a few hundred kilometers from one of the largest metropolitan regions in the world. Dealing with such a crisis in a large urban environment increases the challenges confronting emergency response personnel.

Despite these problems, some aspects of the Japanese response to the 2011 earthquake-tsunami nuclear meltdown did work fairly well. The national, prefecture, and local authorities provided a relatively swift and effective response to the immediate effects of the earthquake and tsunami, especially considering the scope and magnitude of the event and the size of the population affected. More than 100,000 SDFs were on the ground within a week, and food, water, and shelters were quickly provided to hundreds of thousands of individuals.

But other response operations—the removal of debris and repairs to telecommunication and transportation systems—took longer due to political infighting between the LDP and DPJ. This caused public opinion to sour on the overall governmental relief effort, despite the initial positive attitudes expressed during early recovery efforts. Additionally, Prime Minister Kan kept TEPCO management from abandoning the plant, and encouraged plant personnel to continue working to contain the crisis. He accomplished this despite the enormous risks to plant employees and the fact that they and their families were still reeling from the earthquake-tsunami themselves.

The Fukushima nuclear crisis has also forced Japan and many other countries around the globe to reconsider their disaster mitigation and preparedness strategies for nuclear facilities, as well as their reliance on nuclear power. Since the disaster, due to energy shortages and frequent blackout periods, many urban Japanese residents have decided to consume less energy. Sixty percent of the Japanese public report conserving energy even after power has returned (Kingston, 2014). In addition, the Japanese parliament has passed legislation promoting alternative, renewable energy sources to reduce the country's long-term dependence on nuclear power (although its complete phase-out is unlikely) (Al-Badri and Berends, 2013).

The impact of Japan's nuclear crisis, however, did not stop at the country's borders. Rather, public support for nuclear energy in other nations has decreased following the Fukushima disaster. An overwhelming majority of Germans (70 percent) are now against using nuclear energy due to the risks involved (Bulletin of the American Academy of Arts & Sciences, 2012). And French Socialist candidate, François Hollande, was elected President of France based at least in part on his promise to reduce the country's reliance on nuclear power (Al-Badri and Berends, 2013). These shifts in public support have caused political parties traditionally allied with the nuclear power industry to reevaluate their previous positions and present new policies. While the US did not change its nuclear energy policy, the Nuclear Regulatory Commission did conduct a battery of tests and drills at nuclear plants across the country to prepare for similar worst-case disaster scenarios.

The Hurricane Sandy Disaster

Hurricane Sandy struck the US mainland near Brigantine, New Jersey around 07:30 pm on October 29, 2012. It brought extremely heavy rains, strong winds, and record storm surges to the coastal areas of New Jersey and New York. But Sandy's impact was felt across an immense geographic area. It produced catastrophic flooding, severe transportation disruptions, major communication breakdowns, and widespread power outages across a 1,000-mile radius. Hurricane Sandy affected communities in at least 24 states, from the southern tip of Florida to the north-eastern region of Ontario, Canada. In the US alone, over 60 million people were affected by Hurricane Sandy (National Oceanic and Atmospheric Administration, 2012). The huge storm ranks as the largest Atlantic storm on record (National Weather Service, 2012).

Hurricane Sandy was also an extremely destructive storm. More than 650,000 housing units were damaged, and 300,000 businesses properties were affected (AON Benfield, 2013). Some 23,000 people had to evacuate their homes and seek shelter in temporary facilities. Approximately 9.1 million customers lost electrical power, some for extended periods (AON Benfield, 2013). The huge storm was responsible for at least 150 deaths, and it caused damages estimated to exceed $50 billion. As a result, Sandy has the distinction of being one of the deadliest Atlantic storms outside of the Southern states in US history (Blake et al., 2013). It is also the second costliest natural disaster to strike the mainland US; only the losses incurred by Hurricane Katrina in 2005—now estimated at over $108 billion—are greater (Porter, 2013).

Sandy was an unusual storm in several respects. First, it developed fairly late in the hurricane season, especially compared to many other major weather systems. Sandy formed in mid-October as a tropical wave off the west coast of Africa, progressed to a tropical storm by October 22, and intensified to hurricane status on October 24 (National Weather Service, 2012). Sandy also took an unusual path before hitting the US mainland. The storm was traveling in a northerly direction when it struck Jamaica on October 24 and Cuba on October 25. Then, it began moving parallel to the East Coast of the US. However, on October 29, it ran into an "anomalous blocking" weather system that prevented Sandy from following a more typical route (either drifting out to sea or tracking into hurricane-prone areas like Florida and the Carolinas). As a result, Sandy took a surprising hard left turn to the northwest, and it moved directly toward the New Jersey coastline (Gerhardt, 2012). Due to this combination of factors, Sandy's size increased dramatically, expanding to over 1,000 miles in diameter (Gerhardt, 2012).

Sandy will not only be remembered because of its size and impact, but also because of where it occurred. The coastal communities of New Jersey, New York, and Connecticut represent one of the most densely populated and commercially active areas of the United States. As a consequence, Hurricane Sandy had a profound effect on large numbers of people and businesses. Two million New Yorkers (especially those in New York City) and 5 million New Jerseyites lost power (often for several weeks or more). Furthermore, storm surges and flooding damaged 305,000 homes in New York and 340,000 homes in New Jersey (Blake et al., 2013). Similar impacts were felt by the area's business community. Some 480,000 commercial establishments, employing 6.7 million people, were affected (Mesenbourg, 2012). In the states of New Jersey, New York, and Connecticut, total business losses are estimated to exceed $300 billion (Blake et al., 2013). Clearly, Hurricane Sandy had an extremely disruptive impact on the lives of local residents and on the operations of private businesses in the Northeast region of the United States.

There are several other factors that exacerbated Sandy's impact in this area. Not only is this one of the most densely-populated and commercially-developed areas of the US, but it also has some of the most extensive infrastructure and transportation systems in the country, including bridges, dams, tunnels, subways, roadways, railroads, and airports (New York City Panel on Climate Change, 2009). Many of these systems were already under stress and heavily used even before the monster storm hit (Jacob, 2009; Klineberg, 2013). The intense rains, wind, and storm surges further weakened their capabilities, and in some cases led to their collapse or breakdown.

It is also important to point out that residents in this part of the country may simply not have been anticipating the possibility that a major storm like Sandy would hit this region (Bowman, 2013). After all, many (if not most) of these Atlantic weather systems move out to sea or they strike in other places. Hence, many of those living and working in the coastal communities of Connecticut, New Jersey, and New York have never experienced a "direct hit" from a major cyclone like Sandy.

The US Governmental Disaster Response System

The US government has developed a fairly elaborate response framework to deal with major disasters. Similar to the Japanese disaster system, the governmental process works from the "bottom up." It begins at the local level and follows a series of pre-specified steps to the state, and ultimately the national, government based upon the severity of the situation (Schneider, 2011). Over the last 60 years, this basic intergovernmental framework has been used to address thousands of emergency situations in the US. For the vast majority of these events, public institutions have been able to respond in a fairly smooth and effective manner, providing aid and relief to stricken communities. This has been the case even when the response process has been activated to handle extremely large, disruptive situations, such as the 1993 Midwest floods (US Senate, 1994) and the 1994 Northridge earthquake in Los Angeles (US House of Representatives, 1996).

Yet, the governmental response to *other* disaster situations has not always proceeded smoothly or effectively. This was definitely the situation when Hurricane Katrina hit the Gulf Coast of the United States in the summer of 2005. Hurricane Katrina was such an extremely intense and destructive storm that any governmental response, no matter how well coordinated and orchestrated, would probably have encountered serious difficulties. But the problems which surfaced during the governmental response to Hurricane Katrina were not solely due to the magnitude of the disaster. Instead, they are often attributed to breakdowns which occurred within the nation's intergovernmental response system (Carson and MacManus, 2006; Scabo, Kearney, and Kilroy, 2007; Schneider, 2005; 2011; Waugh, 2007). The *Post-Katrina Emergency Management Reform Act of 2006* and the *National Disaster Recovery Framework* were adopted to address the problems that were so evident in the Hurricane Katrina recovery efforts (Federal Emergency Management Agency, 2011; Government of the USA, 2006).

As a result of the problematic intergovernmental response to Hurricane Katrina, changes were made to the US disaster relief process (US Department of Homeland Security, 2009). One of the most significant changes involved the roles and responsibilities of the national government. Specifically, the national government was given the ability to become more directly involved at earlier stages in the process, even before a major disaster strikes (US Department of Homeland Security, 2009). For example, it can deploy personnel, resources, and equipment to an "emerging" situation to help state and local officials respond in anticipation of a major event. Therefore, as a result of post-Katrina policy changes, federal officials do not have to wait for a response to work its way up through the intergovernmental framework before they mobilize and organize necessary resources. The national government, however, is not to take over the response operation, even during a major disaster situation. Instead, federal officials are to work directly with state (and local) officials to assess the severity of the situation, to determine the type of assistance needed, and to provide appropriate assistance in a coordinated, effective manner.

Thus, current US policy specifies both a "bottom-up" and "top-down" process for emergency management and relief. The nation's disaster relief system is still based on the idea that the response should start at the local level, only moving up through the intergovernmental hierarchy if situations exceed the capacities of lower-level jurisdictions (Schneider, 2005; Sylves, 2008).[3] This process is clearly specified in the major operating documents of the nation's emergency management system (the 2009 National Response Framework and the 2011 National Disaster Recovery Framework). Yet, a "top-down" process is also warranted when rare, large-scale events threaten. Here, the national government is to take more aggressive steps and enlarge its role in order to mitigate the effects of catastrophic incidents and mobilize the necessary resources to address such circumstances (Federal Emergency Management Agency, 2009; Fugate, 2009). Hurricane Sandy provided the first real test of this two-pronged intergovernmental response process system.

The Response to Hurricane Sandy

Initially, the US governmental response system seemed to address the situations created by Hurricane Sandy fairly well. Emergency management personnel were monitoring the storm long before it struck the East Coast of the United States. The National Hurricane Center (NHC) began releasing public advisories on Monday, October 22, 2013, indicating that a low-pressure weather system had officially

---

3 Local approaches are, however, significantly shaped by Federal program requirements and funding.

developed into Tropical Storm Sandy. It issued warnings for the first, most likely targets in the US—portions of South Florida and the Florida Keys. The NHC continued to update these warnings over the next few days, expanding the watch area to the Carolinas and then into the Mid-Atlantic states (National Weather Service, 2013).

Several days before the storm hit the East Coast, staff from the Federal Emergency Management Agency (FEMA) began working with personnel in Atlanta, Philadelphia, New York City, and Boston, as well as with officials from other public and private sector organizations, to assess the area's readiness and to identify potential needs. On October 26, FEMA issued an "enhanced" alert for the region, enabling it to proactively provide critical guidance and support. FEMA sent a Management Assistance Team to facilitate state emergency management operations; the US Department of Defense deployed coordinating officers to plan and organize response efforts; and, the US National Guard pre-positioned over 60,000 National Guard personnel along the coastlines in Connecticut, Delaware, Maryland, Massachusetts, New Jersey, New York, Pennsylvania, Virginia, and in the District of Columbia to coordinate with local authorities. Then, on October 27, two days before Sandy struck the US mainland, FEMA activated a National Response Coordination Center, establishing a central point of contact to coordinate the response and to connect federal efforts directly with state-level operations (Federal Emergency Management Agency, 2013b).

Local agencies and personnel also took steps to prepare their communities for the upcoming storm. City officials across the area (New York City and Atlantic City) sent communications out to the general public through a variety of means (major television networks, radio channels, public and private websites, Twitter feeds, newspapers, press conferences) in order to make residents aware of Sandy's path and to encourage them to be prepared for a major event. Based on revised projections from the National Weather Service, 375,000 New York City residents were ordered to leave their homes in advance of the storm, representing only the second general population evacuation in the city's history (New York City, 2013).

State officials along the East Coast began developing pre-landfall emergency procedures, and they issued warnings to encourage citizens to monitor the approaching storm and to be prepared. Connecticut's Governor Daniel Malloy, New Jersey's Governor Chris Christie, and New York's Governor Andrew Cuomo activated portions of their State Emergency Operations Center, mobilizing the assets of state agencies and establishing the means to coordinate state and local operations. All three governors declared states of emergency before Sandy struck the area, giving them the ability to draw upon additional resources and emergency powers if needed (Office of the Governor, State of Connecticut, State of New Jersey, and State of New York, 2012).

By the time Hurricane Sandy hit, President Barack Obama had already signed emergency declarations that authorized FEMA to make federal services available to state, tribal, and local governments in several states along the East coast. On October 30, the day after the storm struck, President Obama issued major disaster declarations making federal financial assistance available to individuals and businesses in the hardest hit counties in Connecticut, New Jersey, and New York (Federal Emergency Management Agency, 2013b).

By this time, the federal government's presence in this disaster situation was quite evident. FEMA personnel were already in the area before Sandy hit, and they quickly established basic emergency response operations. Within a week, FEMA had over 5,000 personnel onsite—including FEMA corps members, search and rescue forces, mobile emergency response workers, and disability integration advisors—to support response efforts (Fugate, 2013). In addition to FEMA, a myriad of other federal agencies were also mobilized to respond. The Army Corps of Engineers had pre-positioned personnel who could provide technical expertise and assess generator requirements. Inspectors from the Nuclear Regulatory Commission were in place to ensure the safety of the area's nuclear power plants. The Department of Defense had established incident support bases in several locations, and stocked these bases with water, meals, blankets, and cots. Likewise, personnel form a number of federal agencies (the US Department of Health and Human Services, the US Department of Housing and Urban Development, and the US Coast Guard) had also been dispatched several days before the storm hit, and they were already ready to help locate stranded citizens, distribute critical supplies, and identify available housing (Hurricane Sandy Rebuilding Task Force, 2013).

The day after Hurricane Sandy struck the area, top federal officials traveled to disaster-stricken communities. Department of Homeland Security Secretary Janet Napolitano surveyed damages in New York and Connecticut (Federal Emergency Management Agency, 2013). President Barack Obama and key FEMA personnel visited some of the severely affected areas along the New Jersey coast with New Jersey Governor Chris Christie (Federal Emergency Management Agency, 2013b). Finally, FEMA Director Craig Fugate announced at this early juncture that "FEMA's priority is to accelerate disaster assistance to the storm survivors" (Rockwell, 2012).

Within the next few days, the federal government took additional steps to get the response moving quickly. President Obama approved emergency restoration work in Connecticut, New Jersey, and New York with the federal government picking up 100 percent of the costs for the first 10 days. The President also ordered the establishment of a national power restoration team to "cut through red tape" and get power restored to storm victims as quickly as possible. During the week following the storm, President Obama held 15 briefings and news conferences related to Hurricane Sandy. The message from the White House was clear: the federal government was to "lean forward" and "cut through red tape" with the end objectives of assisting state and local governments and getting critical resources to victims (Hurricane Sandy Rebuilding Task Force, 2013).

The federal government's presence in this disaster situation is still being felt. During the first two weeks of the response, more than 17,000 federal workers were sent to the most damaged areas and over $400 million had been spent by the federal government on storm-related services. By the end of November 2012, four weeks after the storm's onslaught across the East Coast, over 450,000 people had registered for federal disaster assistance and 4,700 were receiving temporary shelter. On the one-year anniversary of the storm, more than $43 billion in federal funds had been designated to help communities affected by Sandy. These funds went to address some of the most pressing problems, such as removing debris, repairing buildings, and restoring water treatment facilities. But they have also been used to help the area protect itself from future storms—by promoting flood plain management measures, strengthening infrastructures, and creating more resilient back-up power systems (US Senate, 2013).

The federal government has continued to assist the areas most affected by Hurricane Sandy. To date, FEMA has approved over 170,000 applications and $3.7 billion in funds for individual assistance to help victims in Connecticut, New Jersey, and New York repair their homes or find temporary housing. This assistance has also enabled individuals, families, and businesses to obtain emergency services, medical aid, housing support, financial aid, and tax relief. In addition, FEMA has spent over $3.5 billion in public assistance funds to help storm-damaged communities rebuild basic infrastructures, resume essential services, rekindle economic development, and restore natural resources in the three most affected states (Federal Emergency Management Agency, 2014).

But it would be inappropriate to ignore the actions of subnational governments during the response to Hurricane Sandy. Across the region, municipal officials tried to assess the extent of damages and provide critical assistance to those in need. In the largest municipalities, efforts were made to alleviate transportation problems, restore electrical power, provide temporary shelter, reestablish communication systems, and distribute essential supplies and assistance to disaster victims (New York City, 2013).

At the state level, emergency preparedness agencies focused on mobilizing essential resources, coordinating statewide efforts, and connecting with their counterparts across the intergovernmental response framework. The governors of the three most dramatically affected states—Connecticut, New Jersey, and New York—were particularly involved: setting up separate emergency operations centers; issuing continuous alerts and warnings to the public; ordering road closures and travel bans; recommending businesses closures, and; asking that non-essential public employees not report to work (Office of the Governor, State of Connecticut, State of New Jersey, and State of New York, 2012–2014). They also requested and received expedited aid from the federal government (Federal Emergency Management Agency, 2013a).

State-level efforts were strengthened by the activation of the Emergency Management Assistance Compact (EMAC) system. EMAC enables personnel and resources from other, non-affected states to be deployed to an impacted region in order to assist in the response efforts (National Emergency Management Association, 2014). Thirty-seven states deployed personnel and resources to support the Hurricane Sandy response and recovery efforts in Connecticut,

New Jersey, New York, Maryland, Pennsylvania, and Massachusetts. This was one of the most successful, large-scale activations the EMAC system has recorded (National Emergency Management Association, 2014).

Lessons Learned from the Response to Hurricane Sandy: Successes and Failures

Hurricane Sandy has been called a "Superstorm," "Frankenstorn," "Blizzacane," and "Snor-easter." Regardless of the name used to describe it, the storm was clearly a *major* event. Individuals, families, and private organizations could not handle this situation on their own. Governmental resources were needed, and the basic intergovernmental response process was activated to coordinate the actions of local, state, and national authorities. But, the scope and severity of the event necessitated a greater role for the federal government. Therefore, the federal government acted in a more "top-down" fashion, taking proactive steps to jump-start the process and to keep it moving. The government used this dual system throughout much of the response to Hurricane Sandy in order to administer assistance to disaster-stricken individuals and communities.

Many elements of the standard "bottom-up" response worked as intended. Local and state officials concentrated on addressing some of the most immediate and critical problems: providing information to citizens, ensuring public health and safety, addressing transportation breakdowns, and restoring power to affected areas. Then, they moved on to deal with other issues: clearing debris, repairing roadways, and re-establishing communication systems. For the most part, emergency management personnel at the local and state levels seemed to understand what needed to be done, and they tried to focus on working with others to attain these objectives.

The actions of the federal government tended to complement the efforts of municipal and state-level agencies. FEMA coordinated and mobilized the response of a number of critical agencies and marshaled additional resources from the larger, umbrella department—the Department of Homeland Security—in which it is organizationally located. FEMA worked closely with state and local officials to identify the most pressing needs of disaster victims, and it mobilized its own federal personnel and supplies fairly quickly. Thus, the federal government provided the resources and support so that lower levels of government, as well as private and non-profit relief organizations, could perform important emergency response operations.

But the federal government also acted more proactively and directly in this situation than it had in previous events, particularly when compared to its actions during Hurricane Katrina. Major disaster declarations were issued even before Sandy struck the US mainland, federal relief workers were pre-positioned in areas likely to be in hit by the storm, critical emergency resources (water, meals, and blankets) had already been sent to local staging centers where they could be quickly distributed, and public officials were in constant contact with their counterparts at the state and local levels to assess the situation as it continued to evolve. Instead of waiting for lower-level jurisdictions to ask for assistance, federal officials took the initiative to mobilize necessary resources and get supplies to affected areas, in some cases even before they were requested.

Once the storm hit the coast, the federal government continued to be deeply involved in the relief effort. FEMA quickly set in place the process to receive requests for disaster assistance, and it had processed more than 58,000 applications in Connecticut, New York, and New Jersey within a week of the storm's strike (Rockwell, 2012). Federal personnel were onsite providing important emergency response operations such as search and rescue efforts; representatives from the Department of Defense, the Department of Transportation, and the Army Corps of Engineers were in place helping to restore power to affected areas; and National Guard forces were on duty to support state and local interventions with search and rescue efforts, the distribution of emergency supplies, and debris removal (Federal Emergency Management Agency, 2013a).

Federal assistance has also been extremely important in helping affected communities recover from Hurricane Sandy. The federal government has absorbed a large portion of the costs associated with rebuilding storm-damaged homes, businesses, roads, and other infrastructures. It has provided a sizable amount of financial aid to facilitate the development of stronger power systems and communication networks across the region. Moreover, it has funded a number of projects aimed at helping coastal communities prepare for future events—financing flood plain management programs, developing pilot projects to improve the area's energy resiliency, and establishing standards to guide the re-building of homes, businesses, and other infrastructures (Hurricane Sandy Rebuilding Task Force, 2013).

These governmental efforts have contributed to positive outcomes. Within two days of the storm, 80 percent of the subways in New York City were re-opened (Ross and Bindley, 2012). Within a couple of weeks, many families were able to move back into their previously flooded homes, and local jurisdictions were able to resume the delivery of basic public services (Federal Emergency Management Agency, 2012; State of New York, 2013). By the following summer, businesses along the New Jersey coast (Yee, 2013) and the Statute of Liberty in New York City had re-opened for visitors (Allen, 2013).

Despite these accomplishments, however, not all aspects of the intergovernmental disaster process operated quickly, smoothly, or effectively. Some municipalities were overwhelmed with the magnitude of the situation; hence, they were unable to help citizens prepare for and respond to the enormous problems that Sandy produced (Liboiron and Wachsmuth, 2013). Other local units released information to the public that was confusing, even contradictory, in terms of the likely path of the storm and what citizens should do to prepare (Ruderman, 2012). Such actions prevented a number of local governments from performing their pre-assigned roles as "first-responders" in the nation's intergovernmental emergency management framework, and they affected the ability of the public to comprehend what was happening and what they should do in response (National Weather Service, 2013).

Additional problems developed when city and state officials established their own emergency management control centers, separate from the existing governmental agencies and outside the standard operating procedures. This was done so that they could obtain information directly from affected areas and deploy critical resources more swiftly and effectively (New York City, 2013). Yet, the creation of these second command centers weakened the intergovernmental response process. At the municipal level, it interfered with the ability of local jurisdictions to perform their pre-assigned emergency response responsibilities. At the state level, it undermined the state's ability to serve as the central clearinghouse for the mobilization of resources and as the sole point of contact to coordinate activities across the intergovernmental system (Lovett, 2014; Odato 2014).

Further difficulties occurred because of the sheer number of governments involved in the response. By itself, the state of New Jersey has over 500 municipalities representing a population base larger than the size of New York City. Across the entire mid-Atlantic region, there are thousands of municipal governments, hundreds of special-purpose districts, and a number of inter-agency partnerships that span across and within levels of government (Wachsmuth, 2013). All of these governments are tied together by a handful of states and commonwealths, with the federal government providing overarching guidance and direction. The very existence of this intricate web of governmental units makes it difficult to distribute public services in a coordinated manner even in the best of times. Hurricane Sandy, however, inflicted serious damage over an extremely large geographic area, spanning an array of governmental jurisdictions. Thus, the pre-existing web of governmental units created obstacles for a smooth and effective relief effort (Williford, 2012). It also made it easy for public officials at one level of government to blame delays or breakdowns in the process on *other* governmental jurisdictions. The mix of agencies involved in emergency relief during Hurricane Sandy contributed to a "blame game" across and within governmental levels (Zernike, 2013).

The scope of Sandy's impact on the region produced a number of situations that were not addressed quickly, smoothly, or efficiently. Power outages and fuel shortages affected some storm-damaged communities for weeks (New York City, 2013; Rexrode and Dobnik, 2012), and others for several months in some locations (Adler, 2013). Hurricane Sandy caused major disruptions to the communication and transportation systems across the region, making it difficult to remove all of the storm debris blocking the roadways, to get the water pumped out of the flooded subways, and to restore telephone and Internet services (New York City, 2013). Further, over a year after Sandy struck the East Coast, thousands of people were still waiting to hear if they would even be able to receive assistance so that they could rebuild their storm-damaged homes and businesses (Caruso, 2013; Zernike, 2013).

There were also problems addressing the needs and conditions of some of the most vulnerable populations impacted by the storm. Residents of nursing homes and adult care facilities were evacuated to temporary shelters that were not equipped to meet their medical needs, dietary requirements, and physical restrictions (New York City, 2013). Patients in area hospitals were sent to other health care facilities not adequately equipped to handle them; many of these overflow facilities were already stretched to capacity with their own patients (Fink, 2013).

It is also impossible to ignore the political issues that surfaced during the response to Hurricane Sandy. Some of this occurred because of the timing of the disaster: The storm hit the US mainland during the fall of a presidential election year. Therefore, it is probably no surprise that a great deal of media attention was focused on the governmental response, especially the efforts of the federal government. Some commentators suggested that the guidance and leadership provided by the president (as well as the rest of the federal government) influenced the results of the election (McCarthy, 2012). Hurricane Sandy has been called the "October Surprise" and it is said to have bolstered President Obama's popularity and helped him get re-elected (Boerma, 2012). Other reports focused attention on the political partnerships and squabbles that developed during the response—the cooperative relationship that emerged between New Jersey's Republican Governor Chris Christie and a Democratic President, Barack Obama (Ruderman, 2012), and the partisan bickering that occurred between Hoboken Mayor Dawn Zimmer and New Jersey Governor Chris Christie (Wing, 2012a; 2012b).

The political controversies surrounding this disaster have not subsided even though it has been nearly two years since Hurricane Sandy hit the US coast. Charges continue to surface that some state officials made decisions about emergency relief measures based on partisan politics (Kornacki, 2012); public officials at the local level may not have taken the necessary steps to warn residents in the path of the storm (Ruderman, 2012); and the federal government has allocated an excessive, inappropriate amount of money and resources for Hurricane Sandy relief operations (Plumer, 2013). Regardless of the viability of these charges and concerns, they emphasize the political saliency of disaster situations like Hurricane Sandy.

Concluding Thoughts and Recommendations

The two disasters discussed in this chapter provide important insights into the ability of governmental systems to address major urban crisis situations. Both the United States and Japan are considered to be global leaders in disaster preparedness and response. Indeed, many of the emergency management policies and procedures that are used around the world have been developed in these nations. Yet, this does not mean that Japan and the United States are always able to respond to disaster situations in an appropriate or timely manner. Indeed, both nations have encountered crises in the past that have strained their emergency response capabilities. This was certainly the case in the United States following its government's inability to address the problems created by Hurricane Katrina in 2005 (Schneider, 2005) and in Japan following the government's failings in responding to the 1995 Kobe earthquake (Shaw and Goda, 2004).

However, Hurricane Sandy's devastation along the US East Coast in 2012 and the Japanese earthquake-tsunami-nuclear crisis of 2011 have raised additional concerns about the overall effectiveness of governmental response processes. A number of legislative committees, executive offices, bureaucratic agencies, and private groups have investigated the responses to these two disaster situations. Many of the inquiries into the Fukushima nuclear crisis stress TEPCO's inability to address critical problems in a swift or appropriate manner and the government's mishandling of events (see BBC News, 2012; Carpenter, 2012; Funabashi and Kitazawa, 2012; Kingston, 2014; t'Hart, 2013). The vast majority of examinations into the response to Hurricane Sandy emphasize steps that the government should take to further strengthen its emergency response process by improving coordination across levels of government, cooperation between public and private agencies, and the overall capabilities of public agencies involved in emergency preparedness (see Federal Emergency Management Agency, 2013a; US Department of Homeland Security, 2013; US Senate, 2013). Other reports and analyses suggest that broader changes will be needed in the way governmental emergency preparedness systems approach urban planning, land-use management, global warming, and climate change (Carter 2012; Flint 2012; Hurricane Sandy Rebuilding Task Force, 2013; Klinebert, 2013; New York City, 2013).

Particular attention needs to be focused on strategies that will improve disaster resiliency of cities and urban populations. These are intensely debated, controversial issues in contemporary democratic societies. As a result, it has been extremely difficult to reach a consensus on how to make urban areas less vulnerable to the next nuclear mishap or super storm (Flint, 2012; Ladislaw and Kostro, 2013; Sobel, 2013). Yet, these are extremely important issues around the world—more than half of the global population now reside in towns with 2,000 or more people. Even larger segments of the American and Japanese populations (83 percent and 92 percent, respectively) live in urban areas (World Bank, 2014). And, many of these urban communities are geographically located in disaster-prone regions (coastal areas in the path of hurricanes, mountainsides that experience mudslides, neighborhoods built on earthquake faults, and communities located close to nuclear facilities). As a result, more disasters are already occurring in these areas. And the costs—to human lives, government programs, and private companies—associated with helping urban populations recover and rebuild from crisis events continue to climb.

The Fukushima and Hurricane Sandy disasters provide important insights into the ability of contemporary governmental response systems to handle major crises when they strike urban areas. They reveal elements of the emergency preparedness process that work fairly well, like early warning systems and evacuation procedures yet they emphasize that additional measures—more focused communication mechanisms and training programs, longer-range planning projects, stronger public-private partnerships, and a better understanding of the impact of technological developments on environmental conditions—are needed if urban communities are going to be better prepared for future disaster scenarios. The lingering concern is whether there is enough sustained interest and support to undertake the more difficult adjustments that are required to accomplish this. Hopefully, the Hurricane Sandy and Fukushima nuclear events will serve as "wake-up" calls to prompt these discussions and facilitate subsequent changes in disaster management frameworks.

References

Adler, M. (2013). In Lower Manhattan, Sandy still keeping businesses dark. *NPR*, January 23, http://www.npr.org/2013/01/23/170079700/in-lower-manhattan-sandy-still-keeping-businesses-dark.

Al-Badri, D. and G. Berends (2013). *After the Great East Japan Earthquake: Political and Policy Change in Post-Fukushima Japan*. Copenhagen: NIAS Press.

Allen, J. (2013). Statue of Liberty set to reopen after Superstorm Sandy. *Reuters*, July 3, http://www.reuters.com/article/2013/07/03/us-usa-statue-idUSBRE96215R20130703.

AON Benfield (2013). Hurricane Sandy event recap report: Impact forecasting. http://thoughtleadership.aonbenfield.com/Documents/20130514_if_hurricane_sandy_event_recap.pdf.

BBC News (2012). Fukushima report: Key points in nuclear disaster report. *BBC News*.

Blake, E.S., T.B. Kimberlain, R.J. Berg, et al. (2013). Tropical cyclone report: Hurricane Sandy. Washington, DC: US Department of Commerce, National Oceanic and Atmospheric Administration, National Hurricane Center, http://www.nhd.noaa.gov/data/trc/AL182012_Sandy.pdf.

Boerma, L. (2012). Hurricane Sandy: Election 2012's October surprise? Face the Nation. CBS News, October 28, http://www.cbsnews.com/news/hurricane-sandy-election-2012s-october-surprise/.

Bowman, M.J. (2013). Superstorm Sandy—How did it happen and are we prepared for the future? Stony Brook, NY: Stony Brook University, School of Marine and Atmospheric Sciences, http://www.seagrant.sunysb.edu/media/sandy12/UUPInsight-Sandy020113.pdf.

Carpenter, S. (2012). *Japan's Nuclear Crisis: The Routes to Responsibility*. New York: Palgrave Macmillan.

Carson, K. and S. MacManus (2006). Mandates and management challenges in the trenches: The intergovernmental perspective on Homeland Security. *Public Administration Review* 66: 532–536.

Carter, N.T. (2012). Federal involvement in flood response and flood infrastructure repair: Storm Sandy recovery. Washington, DC: Congressional Research Service, October 31.

Caruso, D.B. (2013). Year after superstorm Sandy, thousands struggle to recover. *Lansing State Journal*, October 27, 2A.

Christodouleas, J.P., R.D. Forrest, C.G. Ainsley, et al. (2011). Short-term and long-term health risks of nuclear-power-plant accidents. *New England Journal of Medicine* 364: 2334–2341.

Federal Emergency Management Agency (2009). *National Response Framework: Support Annexes*. Washington, DC: FEMA, http://www.fema.gov/pdf/emergency/nrf/nrf-support-all.pdf.

Federal Emergency Management Agency (2011). *National Disaster Recovery Framework: Strengthening Disaster Recovery for the Nation*. Washington, DC: FEMA.

Federal Emergency Management Agency (2013a). *Hurricane Sandy FEMA After-Action Report*. Washington, DC: FEMA.

Federal Emergency Management Agency (2013b). *Hurricane Sandy Timeline*. Washington, DC: FEMA, http://www.fema.gov/hurricane-sandy-timeline.

Federal Emergency Management Agency (2014). *Disaster Declarations by Year*. Washington, DC: FEMA, http://www.fema.gov/disasters/grid/year.

Fink, S. (2013). Shelter from the storms. *The New York Times*, October 28, A25.

Flint, A. (2012). Will Sandy's legacy loom as large as it should past 2012? *The Atlantic Cities: Place Matters*, December 27, http://www.theatlanticcities.com/politics/2012/12/will-sandys-legacy-loom-large-it-should-past-2012/4259/.

Fugate, C. (2009). Post-Katrina: What it takes to cut the bureaucracy and assume a more rapid response after a catastrophic disaster. Written Statement of FEMA Administrator Craig Fugate for the House Committee on Transportation and Infrastructure, Subcommittee on Economic Development, Public Buildings, and Emergency Management. US House of Representatives, July 27.

Fugate, C. (2013). Year later: Explaining the ongoing recovery from Hurricane Sandy. Written Testimony of FEMA Administrator Craig Fugate for a Hearing of Senate Committee on Homeland Security and Governmental Affairs, Subcommittee on Emergency Management, Intergovernmental Relations, and the District of Columbia, November 6.

Funabashi, Y. and K. Kitazawa (2012). Fukushima in review: A complex disaster, a disastrous response. *Bulletin of the Atomic Scientists* 68: 9–21.

Gerhardt, M.B. (2012). Hurricane Sandy—October 28th–31st, 2012. National Weather Service, Weather Prediction Center, http://www.hpc.ncep.noaa.gov/winter_storm_summaries/event_reviews/2012/Sandy_Oct2012.pdf.

Government of the USA (2006). *S.3721 Post-Katrina Emergency Management Reform Act of 2006*.

Hurricane Sandy Rebuilding Taskforce (2013). Hurricane Sandy rebuilding strategy: Stronger communities, a resilient region. Presented to the President of the United States, http://portal.hud.gov/hudportal/documents/huddoc?id=HSRebuildingStrategy.pdf.

Ikeda, S., T. Sato, and T. Fukuzono (2008). Towards an integrated management framework for emerging disaster risks in Japan. *Natural Hazards* 44: 267–280.

Ishibashi, K. (2004). Status of historical seismology in Japan. *Annals of Geophysics* 47: 339–368.

Kingston, J. (2014). Mismanaging risk and the Fukushima nuclear crisis, in P. Bacon and C. Hobson (eds) *Human Security and Japan's Triple Disaster: Responding to the 2011 Earthquake, Tsunami and Fukushima Nuclear Crisis*. Oxford: Routledge, 39–58.

Klineberg. E. (2013). Adaptation: How can cities be 'climate-proofed?' *The New Yorker*, January 7, 32–37.

Kornacki, S. (2014). Christie Camp held Sandy relief money hostage, mayor alleges. *MSNBC*, January 18, http://www.msnbc.com/msnbc/mayor-christie-camp-held-sandy-money-hostage.

Krauss, E. (2013). Crisis Management, LDP, and DPJ style. *Japanese Journal of Political Science* 14: 177–199.

Ladislaw, S.O., S. Sanok Kostro, and M.A. Walton (2013). *Hurricane Sandy: Evaluating the Response One Year Later*. Washington, DC: Center for Strategic & International Studies, November 4, https://csis.org/publication/hurricane-sandy-evaluating-response-one-year-later.

Liboiron, M. and D. Wachsmuth (2013). The Fantasy of disaster response: Governance and social action during Hurricane Sandy. *Periscope* (open access), http://socialtextjournal.org/periscope_article/the-fantasy-of-disaster-response-governance-and-social-action-during-hurricane-sandy/.

Lipscy, P. and K.E. Kushida (2012). The politics of disaster preparedness: Japan's nuclear plant vulnerability in comparative perspective. Paper presented at the CISAC Conference: "Learning from Fukushima: Improving Nuclear Safety and Security after Accidents," October 15.

Lovett, K. (2014). Gov. Cuomo defends handling of Hurricane Sandy response in wake of critical report. *NY Daily News*, February 12.

McCarthy, R. (2012). Hurricane Sandy may be 'October Surprise' politically. *Boston: Civil Rights Examiner*, http://www.examiner.com/article/hurricane-sandy-may-be-october-surprise-politically.

Mesenbourg, T. (2012). What the Census Bureau can tell you about some of the communities affected by Hurricane Sandy. US Department of Commerce, United States Census, directorsblog.blog.census.gov/2012/11/08.

Mimura, N., K. Yasuhara, S. Kawagoe, et al. (2011). Damage from the great East Japan earthquake and tsunami—A quick report. *Mitigation and Adaptation Strategies for Global Change* 16: 803–18.

Ministry of Defense (Government of Japan) (2012). *Operations of Self-Defense Forces for Defense of Japan and Responses to Diverse Situations, Section 3: Lessons from the Great East Japan Earthquake*. Tokyo: Ministry of Defence, http://www.mod.go.jp/e/publ/w_paper/pdf/2012/30_Part3_Chapter1_Sec3.pdf.

National Emergency Management Association (2013). *The EMAC Response to Hurricane Sandy*. Lexington, KY: National Emergency Management Association, http://www.nemaweb.org/index.php?option=com_content&view=article&id=566:the-emac-response-to-hurricane-sandy&catid=53:emac-news.

National Weather Service (2012). *Event Overview: Hurricane Sandy*. Washington, DC: US Department of Commerce, National Oceanic and Atmospheric Administration, October 29, www.erh.noaa.gov/mhf/EventReview/20121029/20121029.php.

National Weather Service (2013). *Service Assessment: Hurricane/Post-Tropical Cyclone Sandy, October 22–29, 2012*. Washington, DC: US Department of Commerce, National Oceanic and Atmospheric Administration, May 13, http://www.nws.noaa.gov/os/assessments/pdfs/Sandy13.pdf.

New York City (2013). *Hurricane Sandy after Action Report*. New York: Office of the Mayor.

New York City Panel on Climate Change (2009). *Climate Risk Information*. New York: NYCPCC, http://nytelecom.vo.llnwd.net/o15/agencies/planyc2030/pdf/nyc_climate_change_report.pdf.

Nöggerath, J., R.J. Geller, and V.K. Gusiakov (2011). Fukushima: The myth of safety, the reality of geoscience. *Bulletin of the Atomic Scientists* 67: 37–46.

Obara, K., K. Kasahara, S. Hori, and Y. Okada (2005). A densely distributed high-sensitivity seismograph network in Japan: Hi-net by National Research Institute for Earth Science and Disaster Prevention. *Review of Scientific Instruments* 76 (doi: 10.1063/1.1854197).

Odato, J.M. (2014). Report rips state handling of Superstorm Sandy response. *Times Union*, February 10.

OECD (2009). *OECD Reviews of Risk Management Policies: Japan, Large-Scale Floods and Earthquakes*. Paris: OECD.

Office of the Governor (2012–2014). *Governor's Office of Recovery and Rebuilding*. Trenton, NJ: State of New Jersey, http://www.state.nj.us/governor/index.shtml.

Office of the Governor (2012–2014). *Hurricane Sandy Announcements*. Albany, NY: State of New York, https://www.governor.ny.gov/search/node/Hurricane%20Sandy.

Office of the Governor (2012–2014). *Storm Sandy Response and Recovery*. Hartford, CT: State of Connecticut, http://www.governor.ct.gov/malloy/cwp/view.asp?a=11&q=513034.

Ohnishi, T. (2012). The disaster at Japan's Fukushima-Daiichi nuclear power plant after the March 11, 2011 earthquake and tsunami, and the resulting spread of radioisotope contamination. *Radiation Research* 177: 1–14.

Okumura, T., K. Suzuki, A. Fukuda, et al. (1998a). The Tokyo subway sarin attack: Disaster management, part 1: Community emergency response. *Academic Emergency Medicine* 5: 613–617.

Okumura, T., K. Suzuki, A. Fukuda, et al. (1998b). The Tokyo subway sarin attack: Disaster management, part 2: Hospital response. *Academic Emergency Medicine* 5: 618–624.

Okumura, T., K. Suzuki, A. Fukuda, et al. (1998c). The Tokyo subway sarin attack: Disaster management, part 3: National and international responses. *Academic Emergency Medicine* 5: 625–628.

Plumer, B. (2013). The government is spending way more on disaster relief than anybody thought. *The Washington Post*, April 29, http://www.washingtonpost.com/blogs/wonkblog/wp/2013/04/29/the-government-is-spending-way-more-on-disaster-relief-than-anybody-thought/.

Porter, D. (2013). Hurricane Sandy was second-costliest in US history, report shows." *Huffington Post*, http://www.huffingtonpost.com/2013/02/12/hurricane-sandy-second-costliest-in-US-history.

Rexrode, C. and V. Dobnik (2012). Hurricane Sandy: New Jersey, New York still struggle with power outages. *Huffington Post*, November 11, http://www.huffingtonpost.com/2012/11/11/hurricane-sandy-new-jersey-york-power-outage_n_2113856.html.

Rockwell, M. (2012). Government response to Hurricane Sandy detailed. New York: Government Security News, http://www.gsnmagazine.com/article/27741/government_response_hurricane_sandy_detailed.

Ross, J. and K. Bindley (2012). Subways, tunnels, resume partial operation following Hurricane Sandy closures." *Huffington Post*, November 1, http://www.huffingtonpost.com/2012/11/01/hurricane-sandy-subways-tunnels_n_2053260.html.

Ruderman, W. (2012). One storm still hasn't blown over. *The New York Times*, October 31, http://www.nytimes.com/2012/11/01/nyregion/atlantic-city-mayor-is-left-off-obama-and-christie-storm-tour.html?_r=0.

Sakakibara, E. (2003). *Structural Reform in Japan: Breaking the Iron Triangle*. Washington, DC: Brookings Institution Press.

Schneider, S.K. (2005). Administrative breakdowns in the governmental response to Hurricane Katrina. *Public Administration Review* 65: 515–517.

Schneider, S.K. (2011). *Dealing with Disaster: Public Management in Crisis Situations*. (2nd edition). New York: M.E. Sharpe.

Shaw, R., and K. Goda (2004). From disaster to sustainable civil society: The Kobe experience. *Disasters* 28: 16–40.

Shinoda, T. (2013). DPJ's political leadership in response to the Fukushima nuclear accident. *Japanese Journal of Political Science* 14: 243–259.

Shrader-Frechette, K. (2012). Nuclear catastrophe, disaster-related environmental injustice, and Fukushima, Japan: Prima-facie evidence for a Japanese "Katrina." *Environmental Justice* 5: 133–139.

Sobel, A. (2013). What we didn't learn from Superstorm Sandy." *CNN Opinion*, October 28, http://www.cnn.com/2013/10/26/opinion/sobel-superstorm-sandy/.

State of New York (2013). *Hurricane Sandy Response after Action Report: Draft.* Albany, NY: The New York State Division of Homeland Security and Emergency Services.

Sylves, R.T. (2008). *Disaster Policy and Politics: Emergency Management and Homeland Security*. Washington, DC: CQ Press.

t'Hart, P. (2013). After Fukushima: Reflections on risk and institutional learning in an era of mega-crises. *Public Administration* 91: 101–113.

US Department of Homeland Security (2013). *FEMA's Initial Response in New York to Hurricane Sandy.* Washington, DC: Department of Homeland Security, Office of Inspector General, September.

US House of Representatives (1996). Hearing before the Subcommittee on Government Management, Information, and Technology. Committee on Government Reform and Oversight. *The Government Response to the Northridge Earthquake.* 104 Congress, Second Session, January 19.

US Senate (2013). *One Year Later: Examining the Ongoing Recovery from Hurricane Sandy*, Committee on Homeland Security and Governmental Affairs, Subcommittee on Emergency Management, Intergovernmental Relations, and the District of Columbia. 113th Congress, 1st Session, November 6.

Vervaeck, A. and J. Daniell (2012). Japan—366 days after the quake ... 19000 lives lost, 1.2 million buildings damaged, $574 billion. *Earthquake Report*, http://earthquake-report.com/2012/03/10/japan-366-days-after-the-quake-19000-lives-lost-1-2-million-buildings-damaged-574-billion/.

Wachsmuth, D. (2013). How local governments hinder our response to natural disasters. *The Atlantic Cities, Place Matters*, October 28.

Waugh, W.L. Jr. (2007). EMAC, Katrina, and the governors of Louisiana and Mississippi. *Public Administration Review* 67 (Special Supplementary Issue on Administrative Failure in the Wake of Hurricane Katrina): 107–113.

Williford, S. (2012). *Hurricane Sandy Stresses Importance of Intergovernmental Cooperation.* Philadelphia, PA: University of Pennsylvania, Fels Institute of Government, https://www.fels.upenn.edu/news/hurricane-sandy-stresses-importance-intergovernmental-cooperation.

Wing, N. (2012a). Atlantic City Mayor Lorenzo Langford: Chris Christie 'dead wrong' in critique of Sandy prep. *The Huffington Post*, October 30, http://www.huffingtonpost.com/2012/10/30/atlantic-city-mayor-lorenzo-langford_n_2045673.html.

Wing, N. (2012b). Lorenzo Langford, Atlantic City Mayor, wants to confront Chris Christie 'mano-a-mano.' *The Huffington Post*, October 30, http://www.huffingtonpost.com/2012/10/30/lorenzo-langford-chris-christie_n_2042850.html.

World Bank, The. (2014). Urban population (% of total). http://data.worldbank.org/indicator/SP.URB.TOTL.IN.ZS.

Yee, V. (2013). Coney Island opens despite setbacks. *The New York Times*, March 28, A18.

Zernike, K. (2013). Year after Hurricane Sandy, victims contest Christie's status as a savior. *The New York Times*, October 23, http://www.nytimes.com/2013/10/24/nyregion/year-after-hurricane-sandy-angry-victims-contest-christies-standing-as-a-savior.html.

Chapter 4
Urban Disaster Recovery: Considerations for Damage Reduction and Sustainability

Zhila Pooyan

Introduction

Over the past 30 years a number of significant earthquakes have occurred, including the Loma Prieta Earthquake (US) in 1989, the Kobe Earthquake (Japan) in 1995, the Kocaeli Earthquake (Turkey) in 1999, the Bam Earthquake (Iran) in 2003, the Wenchuan Earthquake (China) in 2008, and the Haiti Earthquake in 2010. These and other disaster scenarios provide researchers and policy-makers with examples of both successful and unsuccessful recovery efforts. Each case is unique, reflecting differences in the severity of the event, the way it impacted the built environment and the level of preparedness at the time of the earthquake. Not surprisingly, the post-event situations also differ, reflecting the society's capacity to respond, based on pre-event political, social, administrative and economic conditions.

Pre-event planning of recovery activities is critical. The concept of pre-event recovery planning is built on the recognition that much can be done before an event happens to facilitate post-event recovery and improve outcomes. Recovery planning not only includes decisions and actions taken before a disaster to guide post-event decision making, but it also extends through reconstruction, ensuring that relief and recovery activities align with long-term development goals, address actual needs, and enhance resilience to future disasters (ISDR et al., 2012).

Planning recovery efforts in the post-disaster phase can improve disaster stricken communities while strengthening their capacities toward sustainability. They encompass all phases of the post-disaster process, including emergency response, rehabilitation, restoration, and reconstruction. Throughout the recovery period there is a need for comprehensive and accurate information about current conditions that can be used to inform both short and long range goal setting and decision making.

Communities often try to reestablish themselves in forms similar to pre-disaster patterns; the resulting continuity and familiarity in post-disaster reconstruction enhances psychological recovery (Mileti, 1999). However, reproduction of the previous conditions in the built environment is not always the optimal course of action. In addition to replacement of capital stocks and returning the appearance of the community to normal, reconstruction activities provide an opportunity to promote future economic growth and development (Haas, 1977; Smith, 2011).

An inherent difficulty in dealing with natural disasters is their unpredictability. Even when a disaster can be anticipated with a high degree of certainty (for example, where a fault line has been identified), the precise timing and extent of the next disaster event remain uncertain. This uncertainty makes pre-event preparations and mitigation efforts at best speculative. Consequently, no amount of pre-event planning and preparation will make it possible to completely prevent property damage and disruption of activities.

Planning for disaster recovery is associated with hazard evaluation results and disaster mitigation plans (Topping, 2003). Recovery plans use hazard assessment information, such as probability or intensity of events. The elements of a recovery plan—operations or short-term and long-term policies—should be developed based on disaster mitigation plans. Hazard assessment, disaster mitigation plans, and recovery plans are all elements of a comprehensive plan with overarching development goals. While disaster recovery plans should be formulated pre-disaster, they need to be reviewed or revised post-disaster based on damage assessment results. Once a disaster has occurred, there will never be sufficient information to ensure that the optimal decisions are made at each step of the disaster recovery process. The recovery plan acts as an intermediate step between pre-disaster plans and post-disaster plans that directs the post-disaster activities and policies while revisiting disaster mitigation plans and hazard assessments.

It is important to prepare recovery plans based on damage assessment and evaluation. Such evaluations should be incorporated in reconstruction and redevelopment plans to avoid recreation of the same vulnerability pattern. This indicates that recovery plans should include upgrades to capacities and infrastructures at the local level to increase resiliency at the national level. Thereupon, top-down and bottom-up approaches in the recovery period need to go along with each other to ensure sustainable results (Penuel et al., 2013).

The cases described below illustrate several points. One is the necessity to establish and utilize a broad base for gathering intelligence about current circumstances in the disaster areas and how they may have changed since recovery plans were decided. There is a necessity to view information gathering as a continuing process, comprising multiple feedback loops, rather than as a singular task which can be considered complete at any given point in time.

Damage estimation typically relies on reconnaissance teams that explore the damages and losses of each sector. Damage estimation during the first days post-disaster is necessary for the mobilization of resources. The initial evaluation may be subject to confusion and uncertainty; hence, the evaluation needs to be continued until there is some degree of confidence that the inventories of losses in various sectors are accurate. Loss estimation requires teams consisting of experts with different specialties, supplemented by others with indigenous knowledge, to be comprehensive and precise (Berke, 1996; Oliver-Smith, 2004).

The second point relates to meeting the interim needs of those residents displaced by the disaster event. In particular, recovery efforts must deal with the tension between relocating residents to facilitate rebuilding efforts and the desire of residents to remain close to their usual location to preserve accessibility to employment and social networks. Sometimes relocation of settlements is required for geological or seismological reasons. After the Kocaeli Earthquake in Turkey

Figure 4.1 Permanent public housing, Kobe City
Source: Author's photograph.

Figure 4.2 Open space in redevelopment projects, Kobe City
Source: Author's photograph.

in 1999, for example, some settlements were relocated to less seismically active zones to prevent potential damages from future earthquakes (figures 4.1 and 4.2). There is unlikely to be a single relocation strategy that is suitable for all those whose homes have been damaged or destroyed.

Effective community participation is essential for both of the above aspects of disaster recovery. Recovery plans should be prepared with community participation to reflect community opinions and attract their cooperation during short-term and long-term activities. Community level activities post-disaster span all four phases (Schwab et al., 1998). An example of community participation in disaster recovery is the Hanshin Earthquake in 1995 in Kobe City (Japan), which buried alive 20,000 people, 15,000 of which were rescued by ordinary citizens, while 5,000 were rescued by formal rescue and relief groups that included fire fighters, police, and the Japan ground self-defense force (UNCRD, 1995).

In order to be able to participate in decision making, planning and implementing plans, both in normal times and in emergency situations, local communities need organizational structures through which they can identify their own resources, capabilities, and accordingly develop their own plans for improving their living environments. Such organizational structures include community-based organizations (CBO) created at the district level as an intermediary between local, regional, and national levels.

Finally, recovery efforts must determine the appropriate levels of improvements that should be incorporated during the reconstruction process. Even such obvious 'necessities' as applying modern seismic building codes to rebuilding efforts must be tempered by the higher costs and longer time periods necessary for their implementation. Significant upgrading of community development standards, such as widening streets and providing public parks and open space, are subject to similar concerns. Even though the disaster event may seem to provide an opportunity to enhance community quality, the pressures of time and cost make this difficult.

Recovery Experiences

Two earthquake recovery experiences, Hanshin in Kobe city, Japan (1995) and Bam in Bam City, Iran (2003) are presented respectively to show what policies were adopted and what challenges surfaced during the recovery period.

Hanshin 1995 Earthquake

The Hanshin Earthquake occurred in Kobe City, Japan in the Hyogo Prefecture, and was the largest earthquake in the country since the Fukui Earthquake of 1948. This earthquake caused more than 6,000 deaths, ruined 250,000 buildings, and inflicted 10 trillion yen worth of damages (Shiozaki et al., 2005).

The earthquake struck a region with a rapidly aging population; many senior citizens died due to the collapse of their wooden houses. Among human loss, the percent of people older than 75 years old (about 30 percent) was about 10 times the rate of people less than 40 years old (less than 5 percent). The reasons for the higher percentages of aged people dying were the old wooden houses they lived in coupled with the physical limitations preventing them from escaping the ruins of the houses (UNCRD, 1995).

Another reason for human loss was massive fires due to the earthquake. The cause of the fires was the existence of too many vulnerable, flammable wooden houses in the city. Even buildings with firewalls sustained damages, which reduced their fire resistance (Pooyan, 2004). Consequently, the number of fires was beyond the ability of fire fighters. Further, fire engines could not use city fire hydrants since underground water pipes had been damaged. Water supply was interrupted due to earthquake damage to the electrical infrastructure. There were many traffic obstacles in the city, causing a delay in reaching affected areas. In addition, citizens' initial fire-fighting activities could not be organized smoothly in a post-disaster situation that augmented the number and extent of fires (Shiozaki et al., 2005).

Emergency shelter, temporary housing and reconstruction

After the earthquake many residents of Kobe City had to live in evacuation centers as refugees for a long period of time. Since urban infrastructures had been destroyed or no longer functioned properly, even some people whose houses were not seriously damaged moved to evacuation centers or emergency shelters. Emergency shelters included school auditoriums, gymnasiums, and other facilities that had not been previously been designated as shelters. Consequently, living conditions in shelters were poor. The unsuitable conditions of shelters contributed to the deterioration of the health of elderly people and infants. Some shelters were smaller and less people could take refugee there; in general, smaller shelters provided better living conditions (Shiozaki et al., 2005).

The decision was taken to build temporary houses in suburban areas. This policy sent residents of the city away from their home neighborhoods. In central parts of Kobe City where levels of damage were higher, provision of temporary housing was less than in the suburban wards (Figure 4.3). Most temporary housing was built on the outskirts of reclaimed lands and sites scheduled for future development. Priority assignment of temporary houses was given to the most vulnerable populations, such as aged or physically handicapped people; younger households remained inside the city due to their jobs. As a result, communities were shattered as people who used to live and communicate with their neighbors and enjoy the social life of their neighborhood were relocated, and hence isolated (Shiozaki et al., 2005).

The average size of temporary housing units was about 26.4 m^2; each unit contained two rooms, a kitchen, and a pre-fabricated bath. Temporary houses were better than emergency shelters where people had no privacy, but living in temporary houses for a long period of time was difficult for families with children. The structures were made of thin wooden sheets plated with thin, galvanized iron and there was no soundproofing material to prevent noise from other units being heard. Room temperatures rose to 40°C in the summer and dropped to below zero in winter. Some units installed air conditioners, but most dwellers could not afford the energy cost and had to tolerate extreme temperatures (Shiozaki et al., 2005).

Dwellers in these temporary houses had to take buses into the city for daily needs like grocery. The long distance between temporary housing and the inner city made it difficult to commute frequently. Some people had to live in temporary houses for around five years, although these houses were planned to be used for no more than two years (Shiozaki et al., 2005).

There were some temporary housing sites where the residents supported each other, and with the help of volunteers, created meeting centers to facilitate communication. The administration, however, did not pay attention to such relationships when they moved people into reconstructed public housing afterward. Priority was given to elderly people without considering the social relations within their environments (Shiozaki et al., 2005).

In August 1995 the Three-Year Housing Reconstruction Plan was announced and was reviewed the following year because many people preferred to live in

Figure 4.3 Agra-e-Bam pre-disaster
Source: Author's photograph.

public housing. This was based on the results of research on temporary housing dwellers, which found that people preferred increasing the number of public housing units.

The earthquake destroyed 74,234 houses in the inner city of Kobe, and 66,734 of these units were single-family houses, terraced houses, wooden apartments, or shop houses. It was estimated that 90 percent of lost houses were privately owned or rented. The number of applicants for public housing increased, as people who used to live in private rental housing were forced to apply for public housing since there was no hope that low-rent private housing would be reconstructed. Added to this, lower income people who owned their houses before the quake had to move into public housing as well because they could not rebuild their houses without the availability of compensation for individual property owners. Therefore, building public housing became the main housing strategy (Figure 4.3) (Shiozaki et al., 2005).

Other than public housing, an estimated 66,000 private houses were reconstructed in earthquake-affected areas. The number of planned public housing units was insufficient relative to the demand. In Kobe City the reconstructed public housing met just 60 percent of the demand. Several factors contributed to this: people who lost their homes but did not live in temporary housing were not accounted for; public housing was limited to those whose financial condition was difficult enough to satisfy the usual means test for entering public housing; and about 9,000 households living in temporary houses did not respond to the survey

Figure 4.4 Abra-e-Bam post-disaster
Source: Author's photograph.

and were consequently excluded from consideration (Shiozaki et al., 2005). Most reconstructed public housing was located far from the inner city, disappointing more than 90 percent of people who had lost their houses and were hoping to go back to their original districts. The shortage of reconstructed houses and the geographical mismatch allowed only one in four households to come back to their neighborhoods (Shiozaki et al., 2005).

Urban planning projects
Two months after the earthquake, urban rezoning and redevelopment projects became part of the reconstruction plan. These projects altered the parameters of districts with the aim of constructing public facilities. Roads should have a width of six meters as a rule, and were usually planned as a uniform orthogonal grid. Almost all the housing lots were reallocated and realigned. Due to these projects, owners had to contribute residential land without receiving compensation for widen paths and roads, increased space for parks, and other forms of open spaces (Figure 4.4). Residents opposed these measures, submitting 2,365 written objections to the City of Kobe (Shiozaki et al., 2005). Authorities believed that the measures were necessary to prevent rebuilding houses in the previous manner and that funding for urban planning projects could support a smoother recovery.

However, urban planning measures were carried out forcibly, without considering the opinions of people affected by the disaster. In districts where paths were less than four meters wide, housing reconstruction could not take place. These districts thus needed to be rezoned (Shiozaki et al., 2005).

Other urban planning initiatives came in the form of land readjustment projects (congestion improvement projects); 11 such projects were implemented in different districts of Kobe City. Land readjustment projects were meant to improve the local environment by widening existing narrow streets. Although these projects were again forcibly carried out and residents had no choice in the matter, they positively contributed to post-quake urban planning. In these types of projects, land for widening roads was purchased, hence a more favorable opinion of property owners towards them than in instances where no compensation was offered (Shiozaki et al., 2005).

Community participation in disaster recovery

After the earthquake, community-based organizations called "machizukuri councils" were formed in many districts of Kobe City. More than 100 machizukuri councils were established in Kobe City. The word machizukuri means community development, and the purpose of these councils was to create centers through which authorities and residents could work together to complete reconstruction. The origin of such organizations goes back to the 1980s when residents' participation was encouraged (Shiozaki et al., 2005).

Machizukuri councils attempted to meet the needs of residents and promote restoration. Their activities included: providing information regularly, holding symposiums, and giving advice on rebuilding collaboration among councils. Machizukuri councils regularly issued newsletters (Restoration from Hanshin Earthquake Disaster, 1999).

A 2004 study by Kobe University of machizukuri councils in Nagata Ward (in the western part of Kobe city), which suffered much damage due to the earthquake and subsequent fires, found that these councils were mostly in contact with elderly and disabled people. Younger and active residents did not participate much in these communities' activities. Interviewed residents of Nagata Ward expressed confidence in family members and neighbors as reliable sources of information and assistance in times of emergency. Police, firefighters, and volunteers were not seen in as favorable a light. In general, residents of Nagata Ward had limited awareness of the *machizukuri* councils, their mission, or their activities (Pooyan, 2004).

During Kobe's reconstruction and recovery, *machizukuri* councils were mainly organized in districts where residents obeyed the authorities' mandates for reconstruction. Organizations and residents who were against the urban planning policies were excluded from participation and did not become involved in the community development process (Shiozaki et al., 2005).

Bam 2003 Earthquake

The Bam Earthquake occurred in Bam City, in the Kerman province of southern Iran. This earthquake caused catastrophic damage to Bam City and neighboring villages, with a total population of approximately 142,000. More than 26,000 people were killed, 30,000 injured, and up to 75,000 left homeless. Some 85 percent of housing and infrastructure was destroyed. The earthquake damaged *Arge-e-Bam* (Bam Citadel), a 2,500 years old historical monument located in the north-west part of the city. This large, urban complex/monument was a typical Iranian historical construction employing ancient earthen architecture in an arid environment, and it was one of the main tourist attractions of the Kerman Province. The quake destroyed 70 percent of this huge monument (Pooyan, 2010).

Affected people launched rescue and relief efforts by themselves in the immediate post-quake period. While many people were rescued in this way, these early efforts were hampered by unfamiliarity with proper rescue methods. There were many volunteers in Bam City, but due to a lack of organization and experience many of these people left the city too soon. Some foreign rescue and relief teams came to the city as well. These groups arrived late and could then only contribute to the finding of dead bodies. Some of these groups set up a complicated machinery with limited effectiveness in the context of Bam City. As a result, many rescue teams remained inactive and left the city after a short period. Additionally, rescue and relief infrastructure, such as fire and police stations, had been destroyed.

Figure 4.5 Tents located near destroyed houses
Source: Author's photograph.

Figure 4.6 Temporary prefabricated housing, Bam City
Source: Author's photograph.

Emergency and temporary settlement
Due to severe damage to public buildings and schools, these structures could not be used as shelters, which explains why emergency and temporary housing was provided in the form of tents and prefabricated houses. With many destroyed homes and insufficient temporary housing, people had to live in tents for several months (Khazai et al., 2005). Tents were located near destroyed houses, thus respecting the wish of residents to remain close to relief services provided by the government and to their properties so they could keep an eye on them (Figure 4.7).

Some people moved to the camps created by the Red Crescent, but these camps were mainly occupied by tenants and immigrants from villages surrounding Bam City. Living in tents for a long period came with many difficulties:

- Due to the varying climate in the region with cold winters, hot summers, and desert storms, it was very difficult to keep tents warm or cool.
- Temporary baths and toilets were responsible for unhygienic conditions. Some people pitched tents near schools in order to use their facilities.
- Safety and security for residents living in the tents, especially for tents occupied only by women and children, was an important concern.
- Due to the presence of insects and bacteria, some areas were afflicted by stings and skin disease. The Red Crescent used insecticides.

Figure 4.7 Temporary prefabricated housing, Bam City
Source: Author's photograph.

The other type of temporary housing consisted of prefabricated houses in camps. The majority of temporary houses were constructed in the camps outside of city, but because of residents' and local authorities' objections, the number of residential units in these camps was limited to 10,000. These camps were mainly occupied by people who had come from other places affected by the earthquake. Overall, about 36,000 temporary houses were constructed in Bam City and about 9,000 units were constructed in the camps around the city. More than 3,000 units in the camps stayed vacant, as people did not move into them (Khazai et al., 2005).

Although living in temporary houses was better than living in tents, there were some problems with these houses as well:

- Lack of infrastructure services, such as drinking water, electricity, and telephones.
- Shortage of hygiene facilities like baths and toilets in temporary houses.
- Shortage of educational and welfare facilities such as schools, health centers, libraries, and playgrounds in the camps.
- Shelters ill-suited to the hot and dry climate of Bam City. In the camps there was no green space, and the lack of trees only intensified the situation. Temporary shelters on private land were better because they were located inside gardens with trees whose shadows created a more favorable environment.

- Camp establishment without consideration of social and cultural issues. Common use of baths and toilets, as well as a lack of privacy and protection, were serious matters. People used curtains to separate spaces. In general, a neglect of cultural issues were a source of daily problems in the camps.
- Legal problems over land ownership. Construction permits would be issued only when an ownership document was presented. Because of the death of many owners and the loss of documents in the catastrophe, providing proof of ownership was often impossible.
- Temporary houses located on private land near destroyed houses. This situation caused the secondary problems of obstructed debris removal and delayed the construction of permanent dwellings.
- Long-term use of temporary shelters. It became difficult to evacuate camps since residents had nowhere else to go, in many cases because they could not produce titles for the land they own.
- Disparities in construction type. Some people received tents while others were provided with prefabricated structures as temporary housing, which created a feeling of discrimination.

Reconstruction of Bam City

Bam City reconstruction employed a planning process that was the first of its kind in Iran. The Ministry of Housing and Urban Development contracted the Armanshahr Consulting Company for the construction work (Armanshahr, 2004). In order to protect the natural environment, historical sites, and structures, the plan called for minimal change to the spatial layout of the city. Thus, transportation routes were not to be widened and the city's social and cultural characteristics were to be preserved. Therefore, on site construction was the main strategy employed during reconstruction, and relocations were limited. This approach to reconstruction reflected the will to respect the cultural features and character of Bam City (Behzadfar, 2005). Another important objective was reviving the economy. The agriculture and gardening sectors recovered more quickly because palm and orange trees were not seriously damaged, but the rehabilitation of the business sector required assistance programs targeted at about 4,000 commercial units (Bhzadfar 2005).

There were problems with the reconstruction plan, including:

- Earthquake disaster management and damage rehabilitation did not take sufficient account of socio-cultural issues. Other than building restrictions over and near seismic faults, little consideration was given to the possibility of future earthquakes in the reconstruction process. Indeed, little attention went to issues such as building density, location and the distribution of necessary infrastructure, like traffic routes, hospitals, and rescue-relief stations, for disaster management.
- The time frame for the reconstruction, considering the amount of work to do, was too short (Housing Foundation, 2006).

- Privately owned land located along main traffic routes had to be acquired.

The municipality approved about $50 million for the purchase of such land, but owners complained that expropriation prices were inferior to pre-earthquake market values. After the event, land values decreased in some areas due to the discovery of earthquake faults, and such values went up in areas far from faults. Real estate agencies forecast rising land values over the next few years, which explains the reluctance of land owners to cede possession of their land to the municipality at a low post-earthquake value (Architecture, 2004).

The main planning and policy-making agency was the "Bam Reconstruction Guiding Task Force." Task force members represented the different organizations involved in reconstruction. The main activities of the task force involved: promoting community participation, creating suitable opportunities for scientific and technical institutes to participate in reconstruction, and improving the quality of construction in affected areas (Hassani, 2004).

Four working groups were organized by the task force to facilitate its operations. Their respective purpose was:

- goal setting for, and management of, public aid and charitable contributions;
- goal setting for, and management of, international aids;
- the formulation of a plan and of design for Bam reconstruction;
- organization of the Bam provincial reconstruction task force.

The offices of 16 Housing Foundations from different provinces formed auxiliary task forces (10 in urban areas and 6 in rural areas) to start reconstruction operations, strengthen human resources in the affected areas, and coordinate reconstruction efforts. These task forces were responsible for providing equipment, defining lots for housing reconstruction, issuing building permits, and setting planning and construction standards. The task forces also provided guidance on how to obtain loans and other financial aid, but due to unfamiliarity with the area and its culture, there were communication difficulties impeding the effectiveness of their interventions (Moghimi, 2004).

To protect the city's identity, it was decided that selected consulting companies should locate in Bam City, which is what 33 such firms did (Housing Foundation, 2006). These companies prepared housing plans based on agreed upon architectural and urban planning standards and considered residents' expectations. About 26,000 plans were prepared for residential and commercial units. Some less damaged residential units were evaluated for rehabilitation; about 1,000 residential and 500 commercial units were reinforced (Housing Foundation, 2004). Training courses were held for local construction employees, builders, contractors, and welders by the reconstruction task force (Housing Foundation, 2006).

Community participation

From the beginning, reconstruction authorities and the United Nations Development Program (UNDP) emphasized community participation, protecting local identity, and education to achieve sustainable outcomes, but there was no understanding of how to carry out a participation process among organizations involved in reconstruction. Therefore, there was no active public participation processes during reconstruction.

From the beginning, it was non-local labor that performed all construction activities and many local contractors did not receive their contract cards until several months later. Consequently, there were many unemployed people in the affected region.

Some decision makers had come from other cities and their unfamiliarity with the culture and people's expectations was a problem. Further, the contrast between the relatively comfortable living conditions of the reconstruction taskforce staff and the rudimentary conditions facing people affected by the earthquake was a source of tension, which impaired cooperation with the local population.

Although local residents had rescued their family members and neighbors, after a short period residents became reluctant to participate in reconstruction. Furthermore, many volunteers who had come to the affected areas left before reconstruction had taken place. There are many reasons for this:

- Authorities did not consider community participation and did not assign specific tasks to residents. In fact, authorities did not believe that community participation would stimulate reconstruction.
- Some people were not interested in participating due to their mental stress. Authorities did not pay much attention to such matters.
- The way that rescue and relief teams treated people led them to consider themselves receivers of help, which is a passive mindset. Some rescue teams also disappointed people, causing them to consequently lose interest in participation.
- The distribution of goods and services was not carried out in an equitable way; as a result, many residents felt discriminated against. Such feelings made them disinclined to participate (Pooyan, 2012).

Disaster Recovery: Opportunities and Challenges

There are some opportunities during disaster recovery for reduced damage and improved community resiliency in the face of future hazards. Damage reduction could be achieved through: rebuilding stronger houses, preventing construction in dangerous areas, reducing congestion in urban zones, decentralizing urban infrastructure and services, and creating opportunities for community participation in recovery. One such opportunity is the setting up of community-based organizations, through which authorities and residents cooperate to implement sustainable recovery plans.

According to the recovery experiences of Kobe and Bam City, physical changes were implemented post disaster in a fashion that reduces future potential damages, while being responsive to some objections of the residents and, in other cases, contradicting their preferences. Affected people expected a return to their previous neighborhoods and pre-disaster living conditions.

Every recovery experience offers opportunities for improved possibilities of coping with future disasters and an overall improvement of affected communities, but the possibility of achieving these opportunities hinges on the nature of the process of recovery and its adaptation to the social and economic contexts of stricken communities. Damage reduction efforts relying on the participation of the residents of stricken communities foster a more rapid and consistent return to normalcy and a quicker and fuller resumption of life and activities.

Because the process of recovery, as well as its goals and objectives, are always based on specific circumstances, local capacities, leadership, and participation are essential ingredients in the success of a recovery plan. Any opportunity for damage reduction should be adapted to the indigenous environment.

While stricken communities have to deal with many challenges during disaster recovery, time constraints and financial issues are among the main challenges confronting communities dealing with disaster recovery. It follows that governments must involve numerous actors in the fulfilling of recovery goals to ensure optimal long-term recovery. In such situations, coordination among actors engaged in different tasks is critical to positive recovery outcomes.

It is important to point out that some organizations and actors are involved in recovery tasks temporarily and are not accountable for long-term recovery. Donors, international rescue and relief teams, national governments, and non-governmental organizations provide assistance and contributions at different phases of disaster recovery. The duration of their presence in affected communities and the impacts of their disaster recovery activities in the short-term and the long-term are, however, uncertain. Such matters are most evident in large-scale disasters that have many recovery goals—housing reconstruction, economic revival, and neighborhood improvement—which sometimes compete with each other. Different organizations claim ownership of different aspects of recovery, thus hampering coordination. If long-term recovery is based on local capacities and sources, then outside organizations play peripheral roles, and their departure will not jeopardize disaster recovery.

Considerations in Urban Disaster Recovery

Disaster recovery in cities, especially large cities, is a multidimensional task and a long process. Issues in urban disaster recovery that need to be considered include:

- Recovery plans should be prepared prior to disasters. Recovery is comprised of different phases and each phase has its own requirements. Developing disaster recovery plans prior to a disaster could contribute to disaster mitigation by establishing and maintaining data and inventories. Pre-disaster recovery plans could identify the role of governmental and other organizations, such as community-based organizations. This is important to prevent delays in responses as authorities may not be able to resume their functions immediately in post-disaster situations. Also, pre-disaster recovery plans could facilitate emergency response in subsequent phases.
- Recovery plans should be included as one of the components of urban comprehensive plans. Comprehensive plans in cities have different elements including housing, land-use, transportation, economic development, environmental, and hazard or risk evaluation. Hazards need to be included in comprehensive plans to ensure that related policies are adopted alongside urban development policies. In such a way, hazard considerations are brought into urban policy and risk reduction measures are integrated into urban plans. Hazard evaluation is done through hazard identification, analysis, and development of mitigation and post-disaster recovery planning.
- Recovery must comprise both physical and non-physical aspects like mental health and economic recovery. Recovery is achieved when an affected community rebuilds damaged physical and social structures. Mental scars of disasters can last for decades, and while generally invisible, influence the pace of recovery. Non-physical aspects of recovery like those addressing mental health, social, and economic damages are interrelated; business resumption facilitates mental recovery, and cultural considerations in reconstruction reinforce the social and economic foundations of affected communities.
- Recovery plans should be flexible. They should be capable of change over time as social and economic contexts of affected communities evolve, and unanticipated problems and challenges arise during the recovery period. Emergency response must deal with uncertainty, as situations vary unexpectedly. Long-term recovery and reconstruction are dependent on the economic conditions of affected communities, which are subject to fluctuations. The social environment is also subject to changes, for example, as residents look for new opportunities in other areas or cities. Recovery plans need to be adaptable to unexpected situations.
- Citizen participation is essential in disaster recovery. Recovery plans aim to improve the resiliency of affected communities in confronting potential hazards by relying on their own capacities. This implies that pre-disaster and recovery plans merge with community plans in order to achieve resiliency. Citizen participation may cause short term delays but can contribute to better long term outcomes. Community participation provides education to the public on how to recover from disasters through pre-disaster mitigation planning. Therefore, community-based organizations should involve

the public in risk reduction and post-disaster plans while improving the welfare of neighborhoods in ordinary times. Different groups of residents have different capabilities that could be identified by community-based organizations so they could be mobilized in a post-disaster situation.

References

Andrew, C. and R. Spence (2002). *Earthquake Protection*. London: John Wiley & Sons.
Architecture and Research Planning Research Center (2004). *Proceedings of the Special Workshop on Bam's Sustainable Development*. Tehran: Architecture and Research Planning Research Center.
Armanshahr Consulting Engineering Company (2004). *Comprehensive Plan of Bam City*. Tehran, Armanshahr Consulting Engineering Company.
Behzadfar, M. (2005). *Historical Sites*. Tehran: Abadi Journal, 82–96.
Berke, P.R. (1996). Enhancing plan quality: evaluating the role of state planning mandates for natural hazard mitigation. *Journal of Environmental Planning and Management*, 39(1): 79–96.
Hassani, N. (2004). *The Necessity of Revising Earthquake Disaster Management in Iran*. Teheran: Iran-Japan Workshop, Management and Planning Organization.
Housing Foundation (2006a). *A Brief Report on Bam and Lorestan Reconstruction Rural Housing Rebuilding Plan*. Tehran: Housing Foundation, Office of Reconstruction and Rural Housing.
Housing Foundation (2006b). *Analytical Report on Bam Reconstruction Trend*. Tehran: Housing Foundation.
International Strategy for Disaster Reduction and International Recovery Platform (2012). *Draft Guidance Note on Recovery Pre-Disaster Recovery Planning*. Kobe: International Recovery Platform Secretariat.
Jha, A., K.J. Duyne Barenstein, P.M. Phelps, et al. (2010). *Safer Homes Stronger Communities*. Washington, DC: The World Bank, Global Facility for Disaster Reduction and Recovery.
Khazai, B. and E. Hausler (2005). International shelters in Bam and permanent shelter reconstruction in villages following the 2003 Bam, Iran earthquake. *Earthquake Spectra* 21(S1): S487-S511.
Mileti, D.S. (1999). *Disasters by Design, a Reassessment of Natural Hazards in the United States*. Washington, DC: Joseph Henry Press.
Moghimi, M.H. (2004). *Lessons of Previous Experiences and Future Necessities of Disaster Management*. Tehran: Iran-Japan Workshop, Management and Planning Organization.
Office for the Coordination of Humanitarian Affairs (OCHA) and Shelter Center (2008). *Transitional Settlement and Reconstruction after Natural Disasters*. Geneva: United Nations.

Oliver-Smith, A. (2004). Theorizing vulnerability in a globalized world: A political ecological perspective. *Mapping Vulnerability: Disasters, Development and People.* London: Earthscan, 10–24.

Peneul, B., K.M. Statler and R. Hagen (2013). *Encyclopedia of Crisis Management.* Thousand Oaks, CA: Sage.

Pooyan, Z. (2004). *A Study on Community Disaster Response: A Way toward Vulnerability Reduction.* Doctoral Dissertation, Kobe, Graduate School of Science and Technology, University, Kobe.

Pooyan, Z. (2012). "Earthquake emergency management: Policies for sound emergency response." Lisbon: 15th World Conference on Earthquake Engineering.

Pooyan, Z. and M.R. Ghayamghamian (2010). A study on Bam earthquake reconstruction: An experience in urban scale earthquake reconstruction. Toronto, 9th US and 10th Canadian Conference on Earthquake Engineering.

Restoration from Hanshin Earthquake Disaster (1999). *Key Terminology in Restoration from Hanshin Earthquake Disaster*. Kobe: Supporter's Network for Community Development Machizukuri.

Schwab, J., K.C. Topping, C.C. Eadie, et al. (1998). *Planning for Post-Disaster Recovery and Reconstruction.* Chicago, IL: Federal Emergency Management Agency and American Planning Association, Planning Advisory Service Report Number 483/484.

Shiozaki, Y., E. Nishikawa, and T. Deguchi (2005). *Lessons from the Great Hanshin Earthquake*. Kobe: Hyogo Research Center for Quake Restoration.

Topping, K. (2003). The role of city planning in reducing disasters, *3rd DPRI-IIASA International Symposium on Integrated Disaster Risk Management: Coping with Regional Vulnerability (IDRM-2003), 3-5 July, Kyoto, Japan.*

United Nations Center for Regional Development (1995). *Comprehensive Study of the Great Hanshin Earthquake*. Nagoya: United Nations Center for Regional Development.

United Nations Disaster Relief Coordinator (1982). *Shelter after Disaster*. Geneva: United Nations.

Chapter 5

The Earthquakes in Lisbon (1755) and Angra do Heroísmo (1980): Reconstruction Processes and Actual Risk

Artur Feio, Roberto Bologna, Daniel Monteiro, and Daniel Félix

Introduction

Portugal is considered a moderate-risk country in its vulnerability to earthquakes. However, there have been strong events in the recent history of Portugal, and the earthquakes of 1755 in Lisbon and 1980 in Angra do Heroísmo, Azores, were deadly and destructive. Both earthquakes created critical situations, and the strategies followed to recover present several relevant issues, making them important case studies. The 1755 earthquake is also considered one of the most studied cases in history (Muir-Wood and Mignan, 2009).

Earthquakes are recurring events and, despite the difficulties to estimate intervals between them, it is predicted that similar seismic actions may happen in the future. Lisbon and Angra do Heroísmo are elevated seismic-risk areas, considered by some authors as the highest risk zones in Portugal (Azevedo et al., 2010). Beyond the size of their population, these cities are worthy of consideration for their heritage significance and economic role as important urban centers. Lisbon is the capital of Portugal and Angra do Heroísmo is a World Heritage site. A destructive scenario in Lisbon could cripple the functioning of the entire country; in Angra, it would represent significant heritage losses.

The aim of this work is to understand achievements and failures during past emergency and reconstruction actions, as well as to assess if reconstruction efforts improved preparedness for future earthquakes. A global analysis of the 1755 and 1980 post-quake actions is presented, summarizing the main strategies deployed to cope with the consequences of the earthquakes and needs of the affected communities, and lessons are drawn from what was and was not successful. The work is entirely based in a literature review and presents a general approach to the consequences of, needs during, and solutions proffered after the 1755 and 1980 earthquakes.

The Lisbon and Angra do Heroísmo Earthquakes

The earthquakes of Lisbon (1755) and Angra do Heroísmo (1980) are the events that probably caused the most destruction and deaths in Portugal. Likewise, both events raised certain issues that make them important case studies for seismic science. Despite other relevant earthquakes registered in the country, these examples still remain as the most important, not only because of their consequences but also due to the emergency and reconstruction actions that followed. In this section, both events are analyzed, considering pre-disaster conditions, consequences, and the actions taken after the disasters.

The Lisbon 1755 Earthquake

The city before the disaster

At the time of the disaster, Lisbon was one of the most beautiful European cities, famous for its wealth, with many palaces and churches, and was a city known worldwide for its commercial activity (Dynes, 2003; Fuchs, 2009). It was also one of the largest cities in Europe, after London, Paris, and Naples, with approximately 160,000–200,000 inhabitants (Maxwell, 2005). However, the country of Portugal was perceived as a backwater and irrelevant, while Lisbon was known as the city of the Inquisition, a center of superstition and idolatry, and dominated by the nobility and clergy (Dynes, 2003).

The city was characterized by an irregular and medieval layout, with narrow streets and poor sanitary conditions. Estimations suggest that the city had 20,000 dwellings and that just 13,000 were considered safe for use before the disaster (Oliveira, 2005). Most of the buildings were constructed of dry joint stone masonry with small openings and wooden floors and roofs.

The earthquake

The earthquake struck Lisbon the morning of November 1, 1755, was followed by a tsunami, and finally by fires that lasted between five and six days, a triple disaster (Azevedo, Serrano, and Oliveira, 2009). In spite of the difficulties in establishing the precise magnitude of the earthquake, most studies estimate it at between 8.5 and 9 on the Richter scale (Baptista and Miranda, 2009; Buforn, 2009; Costa et al., 2010; Oliveira, 2008; Zonno et al., 2009), which makes the 1755 earthquake the strongest in Europe (Buforn, 2009; Chester, 2001; Fuchs, 2009). It also has been difficult to determine the number of deaths, since estimates vary considerably, but some sources suggest that 10,000 people died (Chester, 2001). In spite of its high magnitude, the earthquake was not the main cause of these deaths; the tsunami was responsible for much of the loss of life (Chester, 2001), as were the fires that killed those who could not escape due to streets being blocked by debris from collapsed buildings (Oliveira, 2005). The fires also caused huge damages to one third of the city's infrastructure, as buildings were wooden structures that burned after the earthquake. As a result of the destruction wrought by the three events, it is estimated that of the 20,000 dwellings in the city, only 3,000 remained in usable condition after the disaster, and that only approximately 10 percent of the buildings were undamaged, 60 percent had significant damage, and 10–20 percent collapsed (Oliveira, 2005). Beyond the impact of the earthquake, tsunami, and fires, the city was plagued by severe aftershocks (Dynes, 2003); 28 occurred in the first eight days, 250 in the following six months, and roughly 500 by September of 1756 (Oliveira, 2008). Likewise, it is estimated that 10 percent of the national wealth was lost.

The scope of the tri-fold disaster had a great impact in Europe, causing an intense moral and psychological debate. Luminaries such as Voltaire, Rousseau, and Kant have participated in this important debate.

Post-disaster: Emergency and reconstruction actions

The 1755 earthquake is undoubtedly a remarkable event in history, not only because of the magnitude and effects of the triple disaster, but also, and probably most of all, due to its influence on present-day disaster management and reconstruction procedures. It was the first disaster to necessitate a coordinated emergency response and reconstruction program from the state, as well as the first to lead to strategies to reduce future disaster effects (Almeida, 2009; Azevedo, Serrano, and Oliveira, 2009; Chester, 2001; Dynes, 2003; Fuchs, 2009). As a result, it is considered the genesis of disaster management and the beginning of organized research in the fields of seismology and earthquake engineering (Almeida, 2009; Azevedo, Serrano, and Oliveira, 2009; Dynes, 2003; Fuchs, 2009).

These remarkable actions and strategies were taken under the direction of the first Minister of King José I, Sebastião José de Carvalho e Melo, best known as Marquês de Pombal. The actions implemented by Pombal led to a model for subsequent disaster behavior, and many of his strategies are still used today. Pombal also insisted on defining the earthquake as a "natural event," which was one of the most important contentions of the time, as some believed the disasters to be the outcome of divine will, while others saw it as the result of natural causes.

Portugal was the only country going through the state building process, which was common at the time throughout Europe, that had to deal with this scale of destruction in its capital city (Dynes, 2003). The strong and efficient leadership of Pombal reflects one aspect of structural change in the eighteenth century: the development of the modern state (Almeida, 2009).

Emergency response

The tone of the emergency response led by Pombal is best remembered by his answer to the king's question "What should we do?" It is said that Pombal replied, "Bury the dead and feed the living." In fact, Pombal ordered the pursuit of several tasks (Almeida, 2009; Dynes, 2003) which reflected his concern for the survivors: (a) sinking the bodies of the dead in the Tagus River in order to avoid an outbreak of plague (due to the emergency situation, Pombal secured the Church's agreement to disregard traditional religious rites); (b) creating emergency hospitals for the treatment of survivors; (c) implementing measures to assure the supply of food, namely arranging for transport from different food-producing regions, controlling prices, and encouraging fishing; (d) setting up camps to shelter the homeless; (e) using churches as temporary facilities for displaced people, and making clergy responsible for much of the temporary housing and food provision; (f) removing debris and beginning demolition; and (g) adopting legislation to temporarily suspend taxes and mobilize special financial resources.

At the same time, Marquês de Pombal also took action to maintain normalcy by: (a) ensuring public safety by assuring that troops and courts prevent looting and recover stolen goods; (b) controlling land rents and creating laws to forbid the eviction of tenants by landlords; (c) fighting against superstition and prophecies that could scare the population; and (d) keeping the weekly newspaper on schedule to keep the population informed.

While developing and implementing all these actions, Pombal also created an official inquiry and sent it to parochial districts to assess the effects of the disaster (Almeida, 2009; Dynes, 2003). This inquiry contained 13 questions and is considered a landmark in the history of modern seismology, since it gathered the same type of information as is relied upon by contemporary reconstruction processes (Oliveira, 2008).

Reconstruction process

The reconstruction of Lisbon is considered to be as important and innovative as the emergency actions taken in the immediate post-disaster period. According to some authors, the reconstruction was a great achievement in several fields, including urban planning, architectural development, engineering design, and construction. The reconstruction plan included solutions intended to minimize the consequences of futures disasters. One can say that Lisbon's reconstruction was the genesis of risk management.

Rebuilding activity in the city started with demolition and the removal of debris, much of which was then used in rebuilding. Construction on the outskirts of the city was forbidden to avoid dispersing its residents. Then, Pombal asked the chief engineer of the kingdom, Manuel da Maia, to submit a report to guide the reconstruction of the city (Almeida, 2009; Dynes, 2003). He also ordered a survey to register streets, squares, and property boundaries to avoid future problems that could affect reconstruction. The report from Maia recommended controlling the number of floors to decrease damages in the case of collapse and also widening the streets to allow rapid escape despite the presence of debris.

Taking these recommendations into account, a team of architects and engineers prepared a plan considering the myriad issues of structural safety, urban design and aesthetics, public wealth, and building materials, construction techniques, human resources, financial aid, and legal measures such as the control of price speculation. Manuel da Maia noted that allowing a free rebuilding of the city would be too dangerous because people could ignore safety mandates. Any building activity that was not in accordance with the plan was stopped and demolished.

The plan prescribes a regular layout with orthogonal streets adopting a general north-south orientation. This orientation reflects the direction of seismic waves during the earthquake as well as the orientation of the streets that registered less damage (Oliveira, 2008). The blocks of new buildings, known as Pombaline quarter, are characterized by their regularity of measure and shape, creating a visible symmetry and homogeneity. The facades are very similar as well, since architectural details and ornamentation was kept to a minimum. Also characteristic of the Pombaline quarter is the presence of inner yards, introduced to create better natural ventilation, illumination, in a fashion that highlights concern about health and quality of life. Likewise, a water drainage system was introduced, which unfortunately never performed well.

Although all these solutions represented great advances for the time, perhaps the most important improvement was a construction technique called "gaiola" (cage), a wooden structure embedded in the walls, based on the improvement of a wood truss technique already in use in the city at the time of the earthquake. This structure introduced structural flexibility to buildings, which can prevent their collapse in future earthquakes. Its adoption may be due to the fact that timber-framed buildings held up better during the earthquake (Chester, 2001; Oliveira, 2005). As a matter of fact, the King ordered the construction of a wooden palace for himself during reconstruction efforts. A full-scale *gaiola* model was tested for seismic performance using soldiers marching to create dynamic forces. This was one of the first dynamic structural tests of earthquake resistance engineering. Beyond the wooden cage, the rest of the building was developed to perform well against earthquakes, from the foundation to the roof. Moreover, firebreaks were introduced between the roofs of adjacent buildings. All these architectural solutions show how the reconstruction plan was deeply concerned with the possibility of future events, and also how the lessons learned from the disaster significantly influenced the reconstruction process.

Finally, the reconstruction process heeded other important issues. For the first time in Portugal, methods of mass production and prefabrication were used (Chester, 2001; Ramos and Lourenço, 2000) in order to speed construction, guarantee uniformity, and ensure quality control. In the same manner, regulations assured that the materials were produced in the country, which supported local companies and also facilitated transportation and maintained low fabrication prices. All of these factors contributed to an expedited reconstruction process and boosted the local economy.

As we can see, the overall reconstruction plan was innovative in several aspects, and many of these innovations are still commonly used today. While being concerned with all these issues, Pombal was also aware of possible consequences of the earthquake on the economic performance and international status of the country as a whole. Thus, while planning and being responsible for the reconstruction, he was also aware of the importance of the economic viability of Portugal and sent warships to Brazil, India, and Africa to show these lands that trade with Portugal was still secure. He also sent troops to Algarve in order to fight off pirates from Africa, who might try to take advantage of trade opportunities while the country was focused on the reconstruction.

The Angra do Heroísmo 1980 Earthquake

The city before the disaster

Angra do Heroísmo is a city located in the Azores, a group of islands in the Atlantic ocean that are subject to considerable seismic and volcanic activity. It is one of the most important cities of the archipelago; its foundation is linked with the maritime explorations of the fifteenth and sixteenth centuries when the city was an important strategic point. This strategic point is still important today as testified by the presence of NATO at the Lajes Field on the island.

At the time of the disaster, urban and industrial development was not extensive (Oliveira, 1992). The territory was, and still is, characterized by distinctions between urban and rural areas, which extends to construction types. Urban buildings were of better quality, but in general construction quality was primitive and did not, therefore, guard against damage from strong seismic events. Despite this gap, Angra do Heroísmo presented a heritage-rich built environment, a homogeneous urban ensemble, several monuments, and a distinctive vernacular architecture (ICOMOS, 1983) using local materials and traditional construction techniques. The local economy depended on agriculture and cattle, mainly in rural areas where most families had a vegetable garden, while services and commerce prevailed in the city center.

The earthquake

The earthquake struck the island on January 1, 1980 and reached a magnitude of 7.2 on the Richter scale; the event had more impact than any natural disaster in Portugal during the last 200 years (Oliveira, 1992). It caused about 50 deaths, hundreds of injuries, and thousands were left homeless. Angra do Heroísmo, where the level of destruction was extremely high was one of the worse affected areas. Almost all the buildings suffered damages—some authors estimate the destruction at about 80 percent (Andrade, 1992)—and 45 percent of the dwellings were classified as damaged or collapsed (Valente et al., 1992), causing more than 15,000 people to be homeless. Many buildings where vital services were located were unusable and most of the streets were blocked by debris, disrupting transport systems (Valente et al., 1992). The severity of destruction was mainly due to the poor quality of construction. However, most of the buildings that had been repaired in the recent past registered a better resistance, and all reinforced concrete structures held up well. Although the hospital, a reinforced concrete building, suffered damages, its emergency services were not affected (Oliveira, 1992).

Electrical power generation stopped, causing a shutdown of energy supply. Although the water supply network did not suffer significant damage, the lack of power stopped the system, and the treatment of domestic water was compromised for some time. The telecommunications network was also affected, mainly due to power loss. The airport remained functional but there was no available transportation to the city.

Post-disaster: Emergency and reconstruction actions

The consequences of the earthquake were widely visible after the shock. The level of destruction called for a quick response, which the city was not prepared for. Although the solutions and strategies implemented did not match those of Lisbon in terms of scope and innovation, emergency and reconstruction processes have been deemed successful; some locally adopted strategies can provide valuable insight.

Emergency response

In spite of being considered a disaster prone area, the city was not prepared to deal with an event like the 1980 earthquake. There was no service of civil protection or structural emergency plan, and no clear command channel to carry out urgent responses. In order to overcome this gap, representatives of the public, military, civil, and ecclesiastical sectors gathered on the day of the disaster, using the building of the Public Security Police, which then became the command center for all interventions. Despite this lack of preparedness, important actions were quickly performed to restore minimum living conditions. After previous seismic crises in the Azores there was a considerable level of out migration. One of the main goals of the emergency response was to avoid a similar exodus. To this end, some actions were implemented (Valente et al., 1992): (a) homeless people were settled in open spaces near the city center; (b) services and equipment were relocated to less damaged buildings; (c) debris was quickly removed and streets cleaned to reactivate the road network; (d) buildings at risk for collapse were stabilized; and (e) basic infrastructure was restored, if only with temporary solutions.

The homeless were first sheltered in tents, then in public buildings, garages, and schools, as well as in houses of neighbors (Guedes, 1992). Due to climate conditions, authorities soon concluded that tents could not be a lasting solution. Therefore, temporary metallic houses were assembled in urban areas, and in rural zones people were provided with wood which, combined with materials from the damaged houses, allowed self-construction of "wood shacks" near damaged houses. This strategy allowed the urban homeless to maintain their activities in the city such as services and commerce, and also allowed the rural population to pursue their agricultural activities.

Beyond the actions described above, some authors have mentioned the important support provided by the people of Lajes Airfield.

The reconstruction program

The city was not only unprepared to deal with such an emergency, but also faced several difficulties with the reconstruction program. There was no urban plan before the disaster, nor a reconstruction plan for after (Valente et al., 1992). Also, there was a lack of specialized professionals to develop and oversee reconstruction projects. Local technicians were uncertain about the solutions to adopt. Moreover, local materials factories were not prepared to deal with the important demand for construction materials, and there was also a lack of construction workers.

Facing all these constraints, the local government, which was responsible for providing an effective reconstruction program, established the Gabinete de Apoio e Reconstrução—GAR (support and reconstruction office). Strategies implemented by the GAR were mainly based on promoting self-reconstruction; in this way it developed the "self-construction support program." This program was the outcome of special legislation, which provided incentives and favorable technical and economic conditions, intended to stimulate reconstruction. At the same time, construction materials were provided for free to those who had their houses damaged. These measures, together with the population's will to rebuild, can be credited for the large-scale reconstruction effort. In fact, the population of Angra do Heroísmo was actively engaged in this effort. Many people started to repair and rebuild their houses the day after the disaster.

Taking advantage of this popular initiative, the GAR took action to overcome local limitations: (a) the approval of reconstruction projects was kept simple through the emission of provisory licenses that allowed people to start rebuilding immediately; (b) reconstruction work was constantly supervised and each site under repair had continuous control, thus avoiding construction mistakes and making it possible to fix problems quickly. This constant supervision on the field was key to the reconstruction process. It made for ongoing dialogue between experts and residents. The GAR also provided guidelines, support to local designers (Valente et al., 1992), and training for local workmen. The army made a significant contribution as well, helping with the construction of several houses.

The starting point for reconstruction work was the demolition of damaged structures. A substantial effort was undertaken to unblock streets, as well as to control demolition in buildings that threatened to collapse. Technicians checked and verified the security of all damaged buildings to determine their habitability. Most of the debris was reused in several ways: for leveling the soil, rebuilding other buildings, and also as fill to create new spaces along the coast where new leisure areas and social facilities were created.

Although it encountered limitations and difficulties, the reconstruction of Angra do Heroísmo provided an opportunity to plan for the development and growth of the city. Thus, many temporary settlements were provided with the expectation that they would become permanent. Despite some problems, the areas used for temporary housing gradually became permanent new neighborhoods (Valente et al., 1992).

One of the most important, and also successful, achievements of this reconstruction program was the preservation of the historical legacy of the city. Angra do Heroísmo has a significant built heritage that was severely damaged by the earthquake. However, preservation of historical buildings was mandatory during rebuilding. Just three years after the disaster, and while reconstruction work was still ongoing, UNESCO designated the city as a World Heritage site, which endorses the respect that was given to the pre-quake features of the built environment. This attitude toward preservation does not mean that improvements were not considered. Rebuilding included the reinforcement of buildings to mitigate future risks. Generally, the reconstruction program provided an opportunity to improve housing conditions and to increase seismic resistance while maintaining the architectural identity of buildings.

Lessons Learned from Both Events

The two case studies present completely different reconstruction strategies, both of which resulted in positive outcomes. One of the most important lessons is that there is no one solution that fits all circumstances, and different strategies adapted to distinct contexts may be needed to achieve positive results. Another important conclusion is that preparedness involves drawing on prior experience. At the time of the disaster in Angra do Heroísmo, important achievements from the 1755 earthquake in Lisbon were already well known and studied. Unfortunately, the knowledge developed more than 200 years before was not considered for the preparation of Angra do Heroísmo for a future disaster, even though the area is recognized as seismic-prone. Several solutions and strategies developed in Lisbon would have undoubtedly contributed to decreasing damages in Angra do Heroísmo. However, local authorities did not pay attention to risk or to the available resources in the country. Thus, one can conclude that the impact of the 1980 earthquake was not due to lack of knowledge concerning preparedness or seismic resistance, but mainly to insufficient concern for its use by local decision makers.

In fact, these cases demonstrate the tremendous influence of governments in dealing with this kind of event, not only in terms of post-quake actions but also in terms of establishing preventive measures. When the earthquake struck Lisbon in 1755, Portugal, like several other European countries, was going through the state building process. Inherent in the concept of the modern state was the notion of collective responsibility for the welfare of citizens. This may explain why we categorize Lisbon's earthquake as the first modern disaster in which the state accepted responsibility for managing emergency response and spearheading the collective effort for reconstruction.

The plan of Pombal for a completely new part of Lisbon can be interpreted in different ways. First of all, the plan incorporated many lessons learned from the earthquake, thus expressing the aim to be ready for future disasters. This also reflects the commitment of Pombal to define earthquakes as natural events for which society can be prepared, rather than as a divine occurrence. There was also a political intention to the plan. Pombal was trying to build a more economically and politically secure country in the context of a competitive Europe. The Portuguese economy was growing due to profits from the Empire, and Pombal was able to plan for a grand style reconstruction of Lisbon (Dynes, 2003) as an expression of the economic might of the country. The emblematic new square of the plan was, and still is, called Praça do Comércio (the commerce square), which reflects the direction imagined by Pombal for the future of the country. The splendor of the plan demonstrated the will of Pombal, and its power assured that the construction of any unauthorized building, which did not follow the plan, was stopped.

Therefore, the reconstruction of Lisbon reflects two main intentions: (a) better preparedness for future events; and (b) governmental promotion of Portugal in the European context of the eighteenth century.

More than two centuries later, the social, political, and economic context was completely different, as were the reconstruction strategies performed in Angra do Heroísmo. Although seismic science was more advanced and widely known, the city did not learn from previous examples. Despite being a seismic risk area (previous events had happened in other Azores islands), the local government did not implement measures to improve preparedness. Thus, authorities were unable to cope rapidly with such a level of destruction. Since Angra do Heroísmo is an island, it suffers from constraints due to isolation from the main national territory, and this has impaired the availability of resources for reconstruction. While in 1755 Pombal was able to command a wide team of architects and engineers, after the 1980 earthquake, the municipality of Angra do Heroísmo hired an architect for the first time, a first also for the entire region of Azores. Despite the need for large teams of experts to manage reconstruction, the municipal technical service after the disaster comprised only two engineers and one architect. This fact illustrates how the dimensions, importance, and location of a city influence its reconstruction capacity. Lisbon two centuries before had more capacity and availability of human and economic resources than Angra do Heroísmo did in the twentieth century.

Fortunately, two main choices resulted in favorable outcomes for the reconstruction process of Angra do Heroísmo. After the disaster, authorities gave priority to addressing the risk that residents would emigrate rather than rebuild their homes and businesses, and concentrated on preserving the historical character of the city. Incentives to self-construction accelerated reconstruction, while engaging the local population in the rebuilding effort and thereby contributing to maintaining their morale.

The worthiness of the decision to preserve the original characteristics of the city was acknowledged by UNESCO, when it added Angra do Heroísmo to its list of World sites, just three years after the disaster. This distinction had a positive impact on the population's pride and sense of belonging, and constitutes an incentive to the ongoing pursuit of heritage-based planning.

The case of Angra do Heroísmo also shows that, despite the lack of preparedness, decision makers were capable of implementing locally adequate reconstruction actions, avoiding the expected emigration problems, dealing with the lack of resources, and taking advantage of the population's will to rebuild. Moreover, they were also able to use reconstruction as an opportunity to improve buildings and quality of life, as well as to plan for the future development of the city. Therefore, Angra do Heroísmo is a successful example, notwithstanding a complete absence of earthquake preparedness.

Lisbon and Angra do Heroísmo Today

After the events, both cities adopted strategies to improve preparedness. Each one invested during reconstruction in a better seismic performance of buildings, developed and implemented emergency plans (municipal emergency plans), and set up institutions such as those promoting local civil protection, which have improved preparedness. The metropolitan area of Lisbon also has a plan dedicated to responses to seismic risk.

However, it seems that there is still much to do, particularly in Lisbon. After the post-1755 rebuilding phase, construction quality decreased considerably. Moreover, solutions implemented during reconstruction have been abandoned, and new practices replacing the ones adopted during reconstruction reduce seismic performance. Studies have identified the poor seismic resistance of Lisbon's buildings and thus the high risk hanging over its population. Catastrophic scenarios are predicted for Lisbon, while similar studies in Angra do Heroísmo demonstrate that reinforcements introduced during reconstruction provide for better performance and reduced damage if another quake strikes.

Probably due to the passage of a long period bereft of major disaster, Lisbon seems to have forgotten the actual risk and the consequences of an important earthquake, while in Angra do Heroísmo the memory of the destructive event is still vivid in the minds of the residents. Despite several studies and proposals pointing to the need to improve preparedness in Lisbon, no significant actions have been taken. The Portuguese population is not sufficiently aware of the risk, nor are average politicians (Azevedo, Serrano, and Oliveira, 2009), and this is a critical problem since earthquakes are repetitive events.

Conclusions

This chapter presented two cases of post-earthquake recovery and reconstruction that followed distinct strategies. While the difference in time of more than two centuries makes these examples almost incomparable, both do offer important lessons for disaster preparedness strategies today. One can say that Lisbon was a revolution in terms of disaster management, risk mitigation, and seismic science, while Angra do Heroísmo revealed a successful example of population engagement, local preservation, and the use of reconstruction as an opportunity for urban improvement. Despite the fact that they flowed from very different strategies, both reconstruction processes resulted in positive outcomes. In the two cases, political action was crucial to the path of recovery, underscoring the critical role of decision makers in defining reconstruction measures and solutions.

The study also demonstrates that although great achievements took place after 1755, national authorities did not take advantage of them to achieve better preparedness in zones that are at risk. The consequences of the 1980 earthquake would have been considerably mitigated had the knowledge gained from Lisbon had been applied. These examples illustrate how governments have a strong influence on the solutions and strategies implemented in post-disaster reconstruction.

The chapter also demonstrates that the level of development, location and importance of a city all have an important impact on its capacity for reconstruction. The isolation of Angra do Heroísmo and consequent lack of resources also contributed to the low preparedness of the city. On the other hand, the research also makes evident that, in spite of the lack of resources, locally adequate reconstruction strategies may lead to positive results.

Therefore, the analysis of Lisbon and Angra do Heroísmo earthquakes provides two very distinct and valuable approaches to reconstruction after earthquakes, with many lessons to be learned for current disaster preparedness and reconstruction actions.

Acknowledgments

The corresponding author gratefully acknowledges the financial support of Fundação para a Ciência e a Tecnologia, FCT, through grant SFRH/ BD/73853/2010.

References

Almeida, A.B. (2009). The 1755 Lisbon earthquake and the genesis of the risk management concept, in L.A. Mendes-Victor, C.S. Oliveira, J. Azevedo, and A. Ribeiro (eds) *The 1755 Lisbon Earthquake: Revisited.* Berlin: Springer, 147–165.

Andrade, A.R. (1992). Metodologia da intervenção para a recuperação do parque monumental, in C.S. Oliveira, A.R. Lucas, and J.H. Guedes (eds) *10 Anos após o sismo dos Açores de 1 de Janeiro de 1980. Aspectos Técnico-Científicos.* Lisbon: Governo Regional dos Açores, Secretaria Regional da Habitação e Obras Públicas, LNEC, 535–561.

Azevedo, J., L. Guerreiro, R. Bento, et al. (2010). Seismic vulnerability of lifelines in the greater Lisbon area. *Bulletin of Earthquake Engineering* 8: 157–180.

Azevedo, J., S. Serrano, and C.S. Oliveira (2009). The Next 1755—myth and reality: Priorities and actions to develop in case of an earthquake in the Lisbon metropolitan area, in L.A. Mendes-Victor, C.S. Oliveira, J. Azevedo, and A. Ribeiro (eds) *The 1755 Lisbon Earthquake: Revisited.* Berlin: Springer, 559–579.

Baptista, M.A. and J.M. Miranda (2009). Evaluation of the 1755 earthquake source using tsunami modeling, in L.A. Mends-Victor, C.S. Oliveira, and J. Azevedo (eds) *The 1755 Lisbon Earthquake: Revisited.* Berlin: Springer, 425–432.

Barata, M.D. (2009). The Lisbon earthquake of November 1st, 1755: An historical overview of its approach, in L. Mendes-Victor, C.S. Oliveira, J. Azevedo, and A. Ribeiro (eds) *The 1755 Lisbon Earthquake: Revisited.* Berlin: Springer, 25–42.

Bettencourt, M. (1992). Reflexões sobre a política de reconstrução seguida após o sismo de 1980, in C.S. Oliveira, A.R. Lucas, and J.H. Guedes (eds) *10 Anos após o sismo dos Açores de 1 de Janeiro de 1980. Aspectos Técnico-Científicos*. Lisbon: Governo Regional dos Açores, Secretaria Regional da Habitação e Obras Públicas, LNEC, 333–346.

Buforn, E. (2009). Seismotectonics of the Azores-Tunisia region, in L.A. Mendes-Victor, C.S. Oliveira, J. Azevedo, and A. Ribeiro (eds) *The 1755 Lisbon Earthquake: Revisited*. Berlin: Springer, 397–410.

Carvalho, J.F. (1992). Comportamento e reparação das redes de distribuição de água e colectoras de águas residuais domésticas, in C.S. Oliveira, A.R. Lucas, and J.H. Guedes (eds) *10 Anos após o sismo dos Açores de 1 de Janeiro de 1980. Aspectos Técnico-Científicos*. Lisbon: Governo Regional dos Açores, Secretaria Regional da Habitação e Obras Públicas, LNEC, 519–521.

Chester, D.K. (2001). The 1755 Lisbon earthquake. *Progress in Physical Geography* 25: 363–383.

Costa, A.C., M.L. Sousa, A. Carvalho, and E. Coelho. (2010). Evaluation of seismic risk and mitigation strategies for the existing building stock: Application of LNECloss to the metropolitan area of Lisbon. *Bulletin of Earthquake Engineering* 8: 119–134.

Cymbron, J.M. (1992). Acções desenvolvidas pela direcção de serviços de habitação nos primeiros meses a seguir ao sismo de 1.1.80, in C.S. Oliveira, A.R. Lucas, and J.H. Guedes (eds) *10 Anos após o sismo dos Açores de 1 de Janeiro de 1980. Aspectos Técnico-Científicos*. Lisbon: Governo Regional dos Açores, Secretaria Regional da Habitação e Obras Públicas, LNEC, 235–239.

Dias, M. (2010). *RTP/Antena 1 Açores*. http://www.rtp.pt/acores/?article=12294&visual=3&layout=10&tm=7.

Dynes, R.R. (2003). The Lisbon earthquake in 1755 : The first modern disaster. Newark, DE: University of Delaware Disaster Research Center, Preliminary Paper # 333. https://www.google.pt/url?sa=t&rct=j&q=&esrc=s&source=web&cd=2&cad=rja&ved=0CDoQFjAB&url=http%3A%2F%2Fudspace.udel.edu%2Fbitstream%2Fhandle%2F19716%2F294%2FPP%2B333.pdf%3Fsequence%3D1&ei=KoS1UvmQLeWL7AbujYGYDw&usg=AFQjCNFPls-Wm_b4ujunD3U8iDnYq1xt-g&sig2=0.

França, J.-A. (2005). A Reconstrução de Lisboa—política, económica, administração, estética e técnica, in F. Luso-Americana (ed.) *O Grande Terramoto de Lisboa*. Lisbon: FLAD e Público, Vol. 1, 307–328.

Franco, J.D. (1992). Efeito do sismo de 1 de Janeiro de 1980 nas infraestruturas de telecomunicações dos CTT nas ilhas Terceira, graciosa e São Jorge, in C.S. Oliveira, A.R. Lucas, and J.H. Guedes (eds) *10 Anos após o sismo dos Açores de 1 de Janeiro de 1980. Aspectos Técnico-Científicos*. Lisbon: Governo Regional dos Açores, Secretaria Regional da Habitação e Obras Públicas, LNEC, 533–534.

Fuchs, K. (2009). The great earthquakes of Lisbon 1755 and Aceh 2004 shook the world: Seismologists' societal responsibility, in L.A. Mendes-Victor, C.S.

Oliveira, J. Azevedo, and A. Ribeiro (eds) *The 1755 Lisbon Earthquake: Revisited.* Berlin: Springer, 43–64.

Guedes, J.H. (1992). Acções de apoio à reconstrução. Notas sobre o funcionamento do Gabinete de Apoio e Reconstrução, in C.S. Oliveira, A.R. Lucas, and J.H. Guedes (eds) *10 Anos após o sismo dos Açores de 1 de Janeiro de 1980. Aspectos Técnico-Científicos*. Lisboa: Governo Regional dos Açores, Secretaria Regional da Habitação e Obras Públicas, LNEC, 241–261.

ICOMOS (1983). World Heritage List. *International Council on Monuments and Sites*.

Lima, M. (1992). Câmara Municipal de Angra do Heroísmo. O licenciamento e fiscalização das obras de reconstrução no concelho decorrentes do sismo de 1.1.80, in C.S. Oliveira, A.R. Lucas, and J.H. Guedes (eds) *10 Anos após o sismo dos Açores de 1 de Janeiro de 1980. Aspectos Técnico-Científicos.* Lisbon: Governo Regional dos Açores, Secretaria Regional da Habitação e Obras Públicas, LNEC, 45–54.

Lucas, A. and C.S. Oliveira. (1992). Análise de alguns dados apresentados nos relatórios do GAR e de dados da SRHOP sobre a reconstrução, in C.S. Oliveira, A.R. Lucas, and J.H. Guedes (eds) *10 Anos após o sismo dos Açores de 1 de Janeiro de 1980. Aspectos Técnico-Científicos*. Lisbon: Governo Regional dos Açores, Secretaria Regional da Habitação e Obras Públicas, LNEC, 633–666.

Manaças, E. (1992). O sismo de 1 de Janeiro de 1980 visto 43 horas após notas e comentários, in C.S. Oliveira, A.R. Lucas, and J.H. Guedes (eds) *10 Anos após o sismo dos Açores de 1 de Janeiro de 1980. Aspectos Técnico-Científicos*. Lisbon: Governo Regional dos Açores, Secretaria Regional da Habitação e Obras Públicas, LNEC, 223–230.

Maxwell, K. (2005). O terramoto de Lisboa 1755, in *O grande Terramoto de Lisboa*. FLAD e Público, Vol. I, 295–306.

Meireles, J. and R. Camilo (1992). Breves notas sobre a acção do Laboratório de Engenharia Civil no Sismo de 1980, in C.S. Oliveira, A.R. Lucas, and J.H. Guedes (eds) *10 Anos após o sismo dos Açores de 1 de Janeiro de 1980. Aspectos Técnico-Científicos*. Lisboa: Governo Regional dos Açores, Secretaria Regional da Habitação e Obras Públicas, LNEC, 275–278.

Muir-Wood, R. and A. Mignan (2009). A Phenomenological Reconstruction of the November 1st 1755 Earthquake Source, in L.A. Mendes-Victor, C.S. Oliveira, and J. Azevedo (eds) *The 1755 Lisbon Earthquake: Revisited.* Berlin: Springer, 121–146.

Oliveira, C.S. (1992). Algumas considerações sobre o comportamento das edificações com elementos em betão armado, in C.S. Oliveira, A.R. Lucas, and J.H. Guedes, *10 Anos após o sismo dos Açores de 1 de Janeiro de 1980. Aspectos Técnico-Científicos*. Lisbon: Governo Regional dos Açores, Secretaria Regional da Habitação e Obras Públicas, LNEC, 461–480.

Oliveira, C.S. (1992). Preâmbulo, in C.S. Oliveira, A.R. Lucas, and J.H. Guedes (eds) *10 Anos após o sismo dos Açores de 1 de Janeiro de 1980. Aspectos*

Técnico-Científicos. Lisbon: Governo Regional dos Açores, Secretaria Regional da Habitação e Obras Públicas, LNEC, III-IV.

Oliveira, C.S. (2005). Descrição do Terramoto de 1755, sua extensão, causas e efeitos. O sismo. O tsunami. O incêndio, in F. Luso-Americana (ed.) *O Grande Terramoto de Lisboa*. Lisboa: FLAD e Público, Vol. 1, 23–86.

Oliveira, C.S. (2008). Review of the 1755 Lisbon earthquake based on recent analyses of historical observations, in J. Fréchet, M. Meghraoui, and M. Stucchi (eds) *Historical Seismology: Interdisciplinary Studies of Past and Recent Earthquakes*. Berlin: Springer Science+Business Media B.V., 261–300.

Ramos, L. and P. Lourenço. (2000). Análise das Técnicas de Construção Pombalina e Apreciação do Estado de Conservação Estrutural do Quarteirão do Martinho da Arcada. *Engenharia Civil, UM* 7: 35–46.

Silva, V.C. (2005). Sistemas construtivos usados na reconstrução. A "gaiola pombalina." Estudos recentes, in F. Luso-Americana (ed.) *O Grande Terramoto de Lisboa*. Lisbon: FLAD e Público, Vol. 1, 329–374.

Spence, R. (2007). *LESSLOSS Report No. 2007/07: Earthquake Disaster Scenario Predictions and Loss Modelling for Urban Areas*. Pavia, Italy: IUSS Press.

Valente, A.T., F.P Morais, A.V. Bouça, and M.M. Aguiar. (1992). Angra do Heroísmo uma experiência urbanística, in C.S. Oliveira, A.R. Lucas, and J.H. Guedes (eds) *10 Anos após o sismo dos Açores de 1 de Janeiro de 1980. Aspectos Técnico-Científicos*. Lisbon: Governo Regional dos Açores, Secretaria Regional da Habitação e Obras Públicas, LNEC, 25–44.

Veludo, I., P. Teves-Costa, and P.-Y. Bard. (2013). Damage seismic scenarios for Angra do Heroísmo, Azores (Portugal). *Bulletin of Earthquake Engineering* 11: 423–453.

Vieira, J.A. (1992). Reflexos do sismo de 80 na forma urbana de Angra, in C.S. Oliveira, A.R. Lucas, and J.H. Guedes (eds) *10 Anos após o sismo dos Açores de 1 de Janeiro de 1980. Aspectos Técnico-Científicos*. Lisbon: Governo Regional dos Açores, Secretaria Regional da Habitação e Obras Públicas, LNEC, 17–24.

Zonno, G., A. Carvalho, G. Franceschina, et al. (2009). Simulating earthquake scenarios in the European project LESSLOSS: The case of Lisbon, in A.L. Mendes-Victor, C.S. Oliveira, J. Azevedo, and A. Ribeiro (eds) *The 1755 Lisbon Earthquake: Revisited*. Berlin: Springer, 233–243.

Chapter 6

From Disaster to Opportunity: Reviving Urban Function after the Canterbury Earthquakes

Joanne Stevenson, Sonia Giovinazzi, Erica Seville, Charlotte Brown, Yan (Alice) Chang-Richards, and Suzanne Wilkinson

Overview of the Canterbury Earthquake Sequence

During 2010 and 2011, Canterbury, New Zealand experienced two major earthquakes and several significant aftershocks, which changed the region in profound and permanent ways. The earthquake series unfolded with separate faults rupturing at different points in time, creating an uneven landscape of disruption and recovery. Prior to 2010, experts were aware that an estimated 50 percent of Canterbury's seismic hazard was due to unknown faults (Stirling et al., 1999). However, public awareness of the region's seismic risk focused on the known and active Alpine fault about 150 km west of the Canterbury region's largest city, Christchurch. While Canterbury has experienced disruptive earthquakes on other faults throughout its recorded history, before 2010 there had been relatively little seismicity for nearly 100 years.

The first in the series of earthquakes that rocked Christchurch occurred on September 4, 2010 just after 04:30 am This M_w 7.1 earthquake originated from a previously unmapped fault approximately 40 km west of the city at a depth of 10 km. Due to its timing and the fault's location in a relatively rural area, there were few serious injuries and no fatalities attributed to this earthquake. However, the earthquake resulted in high levels of shaking damage and caused thousands of tons of silt to surface as soils liquefied (shown in Figure 6.1), which caused extensive damage to land, buildings, and buried infrastructure (CCC, 2011). The damage from this event cost an estimated \$5 billion[1] (Bascand, 2011; Eidinger, Tang, and Rourke, 2010b).

Approximately six months after the September 2010 earthquake, as recovery was starting to gain momentum, a separate fault ruptured leading to a much different set of effects. On Tuesday February 22, 2011, two relatively shallow and intense (M_w 6.3 and M_w 6.1 respectively) earthquakes occurred on a fault under the

1 All figures in this chapter are reported in New Zealand dollars: NZ\$5 billion is about \$3.65 billion US dollars at September 2010 conversion rates.

Figure 6.1 Liquefaction silt and sands
Source: Photograph by Matthew Hughes.

Figure 6.2 Damage to unreinforced masonry
Source: Photograph by Matthew Hughes.

Port Hills approximately 10 km east of the Christchurch central business district (CBD). This earthquake again triggered widespread liquefaction, especially in central and eastern Christchurch, and damaged an estimated 46 percent of greater Christchurch's road, water, and sewage infrastructure (CERA, 2012a). This event also resulted in substantial rockfalls and landslides in the Port Hills to the south and east of the city, and high intensity shaking caused extensive building damage throughout Canterbury. Unlike in the case of the September earthquake, there were 185 fatalities in February 2011, the majority of which occurred in two multi-story building collapses in the Christchurch CBD (NZ Police, 2012).

These earthquakes were not isolated events. They were accompanied by seemingly endless aftershocks, including a swarm of more than 32 shallow earthquakes (the largest of which was M_w 4.9) the day after Christmas in 2010 on a fault almost directly below the Christchurch CBD and an M_w 6.4 aftershock on June 13, 2011 that caused 46 serious injuries and further damage to property (GNS Science, 2011). This series of earthquakes in 2010 and 2011 were the most expensive disasters New Zealand has faced in its modern history. The cost of recovery and reconstruction has been estimated at about $40 billion,[2] equivalent to around 19 percent of New Zealand's GDP (NZ Treasury, 2013).

As a result of the earthquakes, Christchurch experienced moderate population and employment losses. At the time of the 2013 census, the residential population was down 2 percent, with 341,000 people living in Christchurch city compared to 348,000 in 2006. The number of paid employees in the city also fell about 1.2 percent between 2006 and 2013. At the time of writing in 2014, the affected built, social, and economic environments are at various stages of recovery and reconstruction.

The earthquakes catalyzed social and institutional responses that had major implications for the people and organizations pursuing recovery in the city. These responses were informed by past trends and policies, immediate and emerging realities, and by hopes for a new resilient future. Christchurch not only needs to overcome challenges and reduce vulnerabilities created by the earthquakes, but also needs to address, recapture, and reverse trends and issues that pre-dated the disaster. Some post-earthquake urban planning, for example, has attempted to reverse the pre-earthquake hollowing-out of Christchurch's city center that occurred as suburban retail areas flourished on the city's periphery.

In practice, recovery progresses through the decisions and actions of thousands of individuals, households, private organizations, and public institutions. This chapter, however, focuses specifically on legal policy responses and the key players involved in strategic governance following the earthquakes. Some of the most central players at the various levels of government are shown in Figure 6.3. These players include existing entities that had established hazard management and emergency response functions (for example, the regional council Environment Canterbury [ECan][3]), existing entities that adopted new earthquake specific roles, and entities that were created after the earthquakes to coordinate the unprecedented recovery and reconstruction efforts.

The Earthquake Commission (EQC), for example, was established in 1993 to provide insurance for residential property and contents against earthquakes, landslips, volcanic eruptions, hydrothermal activity, and tsunami, and is funding a significant portion of repair and reconstruction work in Canterbury (EQC, 2013). The Ministry of Business, Innovation, and Employment (MBIE), on the other hand, was not engaged directly in hazard management before the earthquakes, but played a key role in the development of several important policies in the earthquake aftermath. Similarly, the New Zealand Transport Agency (NZTA) became a major source of funding and recovery and reconstruction planning after the earthquakes.

---

2 These costs are distributed between residential property ($18 billion), commercial and social assets ($15 billion), and infrastructure ($5 billion) (NZ Treasury, 2013). Social assets include publicly owned buildings and property, for example, schools, libraries, and hospitals.

3 ECan is responsible for environmental resource management and emergency management in Canterbury.

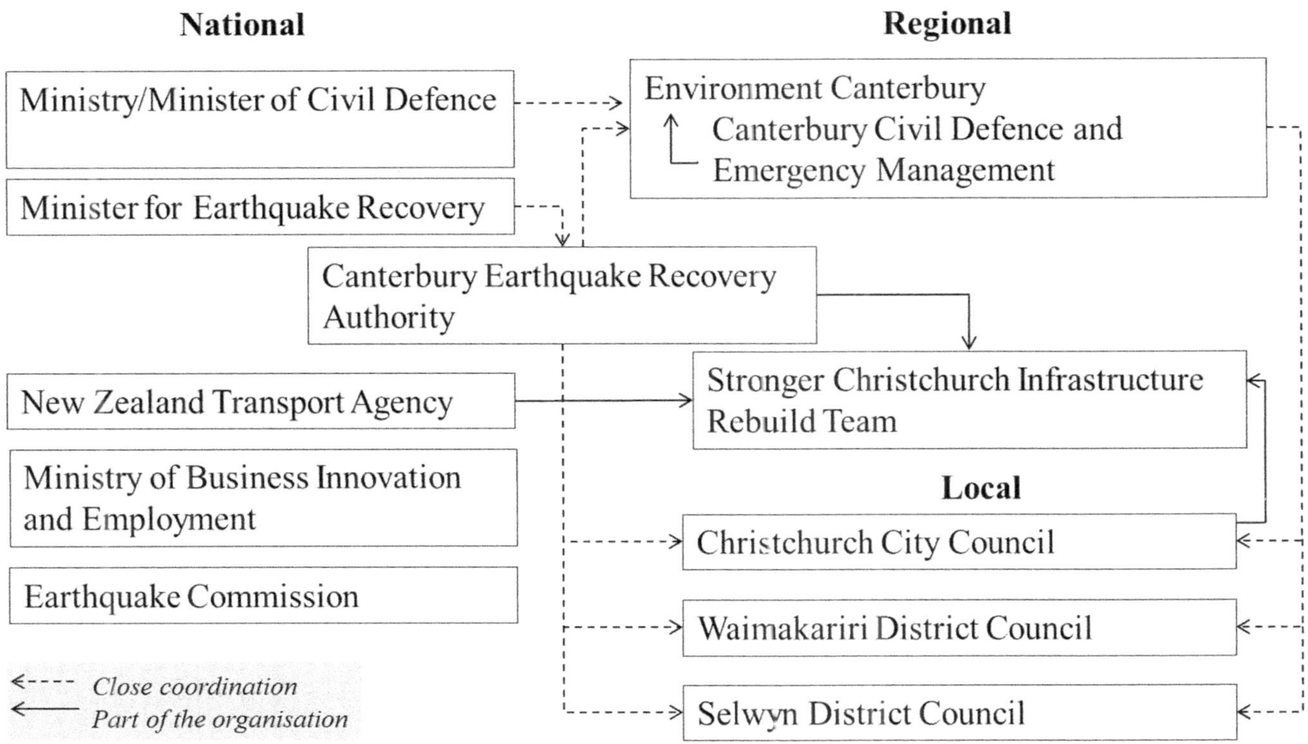

Figure 6.3 **Policy decision makers at various levels of government**
Source: Authors.

As is often the case following major disasters, existing institutions were not adequate to deal with the task of large-scale urban recovery in Christchurch. Therefore, new entities were created in the earthquake aftermath. The position of Minister for Earthquake Recovery, the Canterbury Earthquake Recovery Authority (CERA), and the Stronger Christchurch Infrastructure Rebuild Team (SCIRT) did not exist prior to the earthquakes, but played significant roles in shaping response and recovery. The Minister for Canterbury Earthquake Recovery was appointed less than two weeks after the September 2010 earthquake, when Parliament passed the Canterbury Earthquake Response and Recovery (CERR) Act. The CERR Act facilitated central government response and recovery activities in Canterbury. The Minister was granted special powers to coordinate the recovery and reconstruction activities on behalf of the central government (CERA, 2013).

Following the February 2011 earthquake, the CERR Act was simultaneously repealed and replaced with the passage of Canterbury Earthquake Recovery Act in April 2011. The CER Act 2011 created CERA, a public service department established to coordinate recovery activities in the region. CERA represented New Zealand's central Government on the ground in Canterbury throughout the recovery.

Similarly, SCIRT is a key player that was established in the earthquake aftermath. It is a project management alliance between infrastructural asset owners (including CERA, Christchurch City Council, and the NZTA) and five implementation teams from large construction firms. SCIRT has the formidable task of repairing and replacing horizontal infrastructure (for example, water, sewerage, roads) following the quakes, and is discussed further below.

Response and recovery has evolved considerably since the initial event in 2010, and the future of Canterbury has been progressively reshaped by the decisions and actions of government entities. The combination of established, modified, and new entities to coordinate the Canterbury earthquakes demonstrates the need for planning and mitigation as well as adaptation and innovation to manage complex urban recovery processes.

As of mid-2014, the reconstruction process in Christchurch is only just beginning in earnest. The future of the city is taking shape, but is still characterized by a palpable sense of uncertainty. In this chapter we explore some of the post-earthquake decisions that are having significant implications for Christchurch's hoped for revival as a restored and newly viable and resilient urban area. We examine the implications of post-earthquake land zoning, the program of lifeline utility enhancements, new policies in the construction and insurance sectors, and the strategies adopted to mitigate the economic impacts of the earthquakes in Christchurch. In each section we provide a 'reality check' to evaluate how the events that unfolded after the decision compare against pre-earthquake planning and preparation. We conclude by examining the influence of these decisions on Christchurch's future and by reflecting on the lessons that can be extrapolated for effective urban disaster recovery in other places.

Post-Disaster Land Zoning

The city of Christchurch was planned and built on the Canterbury plains, which were formed by alluvial and fluvial deposits from two overlapping glacier-fed river systems. The same processes that made the area mostly flat and relatively fertile and therefore appealing for European settlement in the 1840s, also made Christchurch vulnerable to significant land damage following the 2010/2011 Canterbury earthquake sequence. Many areas of the city were built on historic river channels, low-lying floodplains, and drained swamps, and later protected from regular flooding by stopbanks and river realignment projects. During the earthquakes, areas such as these were especially vulnerable to ground failure caused by liquefaction, settlement, and lateral spreading. Areas in the Port Hills to the south and east of the city, were vulnerable to rock falls and landslips (Allen et al., 2011).

In response to these land failures, ongoing safety concerns, and the threat of future disruption, CERA made three major zoning decisions. These decisions have significantly influenced the way recovery has progressed in Christchurch and throughout Canterbury. First, the decision to cordon and then declare the Christchurch Central Business District (CBD) a "Red Zone" restricted access to Christchurch's largest commercial centre. Second, CERA decided to pursue comprehensive land zoning and buy out damaged residential properties. Third, in 2014 the Christchurch City Council decided to reassess the flood risk in Canterbury and expand flood management areas on land with increased exposure to flooding as a result of the earthquakes.

Prior to the earthquakes, the regional and local councils were aware of the major hazards that ultimately shaped post-earthquake land zoning decisions: liquefaction, rockfall, and flooding. Reports prepared by The Ministry for the Environment in 1977 and the Christchurch Engineering Lifelines Group in 1997 identified areas that had the potential for extensive liquefaction and acknowledged the dangers of seismically induced rockfalls and rockslides in the Port Hills (CAENZ, 1997; Platt, 2012; St. Clair and McMahon, 2011). Additionally, in 2007, the regional council, ECan, published a liquefaction potential hazard map for Christchurch (ECan, 2007).

As a result of these investigations, more recent building developments required ground improvement activities to mitigate the effects of potential liquefaction. For example, 10 years prior to the 2010 earthquake, renovations to AMI Stadium in the Christchurch CBD included ground improvement work. Stone columns extending 8–9 meters into the ground, reduced the potential for structural damage and improved public safety by maintaining bearing capacity in the stands in the event of earthquake induced liquefaction (Wotherspoon et al., 2014). Compared to liquefaction mitigation, efforts to address rockfall risk had been fairly minimal. In their 1997 reported they noted that, "Some protection measures can be implemented for house sites … but prediction of specific problems or failure sites is not considered generally feasible" (CAENZ, 1997: 271).

The third hazard, flooding, was part of Canterbury's public consciousness and policy since the early European settlement. However, until the 1980s and 1990s flood mitigation measures consisted almost exclusively of engineered structures, such as stopbanks, to protect property and lives. Beginning in the 1980s, the regional council prepared Floodplain Management Plans (FMP) for vulnerable areas, integrating non-structural mitigation measures such as zoning and base elevations for construction to mitigate localized flood risks (CAENZ, 1997; ECan, 2014).

Unfortunately, much of the land zoning in Christchurch and surrounding areas occurred prior to initial liquefaction studies and the development of FMPs. Similarly, the majority of residential and commercial construction occurred before any technical guidance on liquefaction, rockfall, and reassessed flood zones became widely available in the 1990s (Platt, 2012). Even after these risks were better understood, they were not widely appreciated in planning and construction contexts, demonstrated, for example, by the large amount of recent buildings constructed with slab on grade floors (Rogers et al., 2014: 494). This type of construction is especially vulnerable to land deformation and inundation and has been a major issue following other disasters, including Hurricane Katrina (Olshansky and Johnson, 2010).

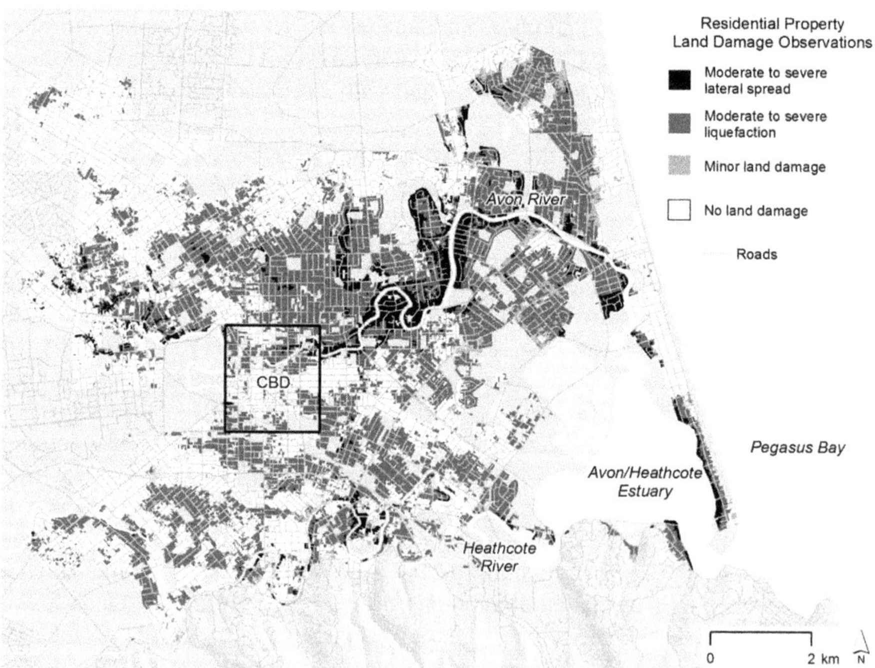

Figure 6.4 Residential property land damage
Source: Authors.

The damage caused by the Canterbury earthquake sequence confirmed earlier hazard assessments. The September 2010 earthquake caused widespread liquefaction, leading to differential ground settlement and land damage throughout the region. Less than a year later, the February 2011 earthquake caused further liquefaction and extensive rockfalls and landslides in the Port Hills area. Large aftershocks including the M_w 6.4 earthquake on June 13, 2011 again caused liquefaction in areas that had already been affected and others that had not previously experienced liquefaction. Figure 6.4 indicates the location and severity of land damage in the Christchurch area.

Liquefaction covered streets, sometimes several feet deep, and spread through people's homes and properties. It not only required extensive cleanup efforts and damaged structures on affected land, but also increased the future risk of liquefaction, as ground settlement thinned the crust. Christchurch also faced increased future flood risk, as liquefaction, associated ground settlement, and construction activity changed the land, rivers, and estuaries (Allen et al., 2011; 2014). In addition to land damage, there was significant shaking damage to buildings throughout Christchurch, especially unreinforced masonry (URM) buildings (Figure 6.2), one of the highest concentration of which were in the Christchurch CBD.

CBD Red Zone

Following each of the major earthquakes in 2010 and 2011, authorities erected a fence and barriers around significant portions of the CBD. These barriers became universally known as the 'cordon' and the area contained within the cordon was referred to as the CBD Red Zone. The cordon was initiated through an executive decision by the National Civil Defense controller following the September 2010 earthquake to regulate access to the city while buildings were searched and assessed for danger. This decision set a precedent for subsequent earthquakes in the months to follow. The cordon was in place for approximately one week following the September 2010 earthquake, and erected again briefly after the 2010 Boxing Day aftershock. After the February 2011 earthquake, a cordon was again established around almost the entire area within the four avenues which define the circumference of the CBD. Parts of the cordon remained in place for nearly 2.5 years after the February earthquake.

The decision to cordon the CBD was officially characterized as a safety measure due to widespread building and infrastructure damage. The destruction in the CBD was, indeed, extensive. Nearly 1,600 commercial buildings (an estimated 70 percent of the total) will have been totally or partially demolished at the end of the Christchurch CBD's reconstruction (Brownlee, 2012). However, the cordon also served a strategic purpose. It eased search and rescue activities and demolition management, making it faster and cheaper to proceed with a large number of demolitions. With the cordon CERA could achieve more comprehensive oversight to, for example, ensure that all demolition teams were adequately qualified (Chang et al., 2014). Additionally, building occupants were not required to pay rent or pursue repairs while their buildings were cordoned, and businesses often qualified for interruption insurance if they could not access their building. Without the cordon there may have been pressure for tenants to reoccupy buildings in areas that were not conducive to work or safety (Stevenson, 2014).

The CBD cordon was unprecedented in New Zealand and internationally in its scale and duration (Chang et al., 2014). In the short- to medium-term, the damage and the cordon contributed to the exodus of businesses and residences from the CBD. In 2013 there were 2,200 fewer businesses in the CBD than in 2010 (a 38 percent reduction) and over 2,700 fewer residents than in 2006 (a loss of about 36 percent of the 2006 population) (Statistics New Zealand, 2013). Long-term, the cordon enabled the centralized demolition and urban planning process managed by CERA. This minimized *ad hoc* reconstruction activity in the years following the earthquakes. Thus, decision makers generally felt that the cordon's contribution to safety, demolition, and planning management outweighed the potential cost of exacerbated displacement and extended detachment of people and businesses from cordoned areas.

Residential Land Zoning

The second major zoning decision was the process of classifying and purchasing residential land. After the February 2011 earthquake, it became apparent that it would not be economically or technically feasible to rebuild in some areas. Changes in geological landscape meant that some areas were more exposed to liquefaction, flooding, and landslides than before. In these areas, even where it was possible to remediate the land to reduce these risks, it was financially unpalatable due to uncertainty in the Christchurch land market, both in terms of land values and future insurability in an environment of heightened risk (Taylor, 2013). Additionally, the Government had an interest in reducing their insurance exposure through the EQC hazard insurance program. The Government and city council were also resistant to repairing and replacing infrastructure in areas with reduced populations and a greatly increased threat of damage.

In an attempt to create certainty, confidence, and simplicity for affected residents, spur the pace of residential recovery, and reduce the Government's risk exposure, on June 22, 2011 CERA announced a new system of land zoning. Initially residential land was zoned into four categories: white, orange, red, and green (described in Table 6.1).

Table 6.1 Categories of land damage

Land zone	Description
White	Land classified as white meant that complex geotechnical issues relating to land slip and rock roll required further assessment and observation before land decisions could be made.
Orange	Land classified as orange meant that engineers needed to undertake further investigation.
Red	Area-wide land and infrastructure damage. Engineering solutions to repair the land would be uncertain, costly, and is likely to be highly disruptive. Land remediation is not considered viable and infrastructure would be difficult and costly to maintain.
Green	Green zone areas are generally considered to be suitable for residential construction.

Source: CERA (2012c).

By the end of 2012 all areas were rezoned to either red or green, eliminating the white and orange categories. As part of this process, the Government through CERA has bought-out over 7,300 residential properties in the Residential Red Zone (CERA, 2014). Homeowners in the residential red zone could either accept: 1) the purchase price paid for the property based on the 2007 or 2008 rating valuation (depending on the area) and the Central Government would take over all insurance claims for the property, building and fixtures; or 2) the Government would pay the recent ratable value and take over insurance claims for the land only, and owners would retain the benefit of insurance claims for the buildings and fixtures. A third option was to refuse any buy-out, but homeowners were warned that the Government would not install new utility services in red zoned areas, and insurers, including the EQC, may cancel or refuse to renew policies in these areas (Platt, 2012; Rogers et al., 2014).

In areas that were zoned green, or suitable for residential repairs and construction, the land was further subdivided into technical categories. These technical categories (Table 6.2) specified the level of geotechnical investigation and foundation design that were needed to address issues with land damage and the risk of future liquefaction.

Table 6.2 Technical categories for repair and rebuilding

Technical category	Description
Grey (TC1)	Future land damage from liquefaction is unlikely. You can use standard foundations for concrete slabs or timber floors
Yellow (TC2)	Minor to moderate land damage from liquefaction is possible in future significant earthquakes. You can use standard timber piled foundations for houses with lightweight cladding and roofing and suspended timber floors or enhanced concrete foundations.
Blue (TC3)	Moderate to significant land damage from liquefaction is possible in future large earthquakes. Site-specific geotechnical investigation and specific engineering foundation design is required.

Source: CERA (2012d).

The zoning and buy-out process has not been without its drawbacks. Residential red zones have fragmented neighborhoods and disrupted social networks. In some cases the 2007/2008 ratable value did not account for improvements and appreciation to people's homes, resulting in financial losses for the homeowners. However, the alternative was stagnation of the land market and enduring uncertainty for people in heavily damaged areas. Zoning provided some sense of certainty where land damage and the risk of further liquefaction damage had effectively shut down the property market in Christchurch. Without a clear sense of liability it was difficult for insurance companies to settle claims and determine where new policies could be issued. Zoning was essential, not only for stabilizing the property market, but the assignment of technical categories also helped focus limited geotechnical and engineering resources on high priority areas.

In the aftermath of the earthquake, it may have been possible to allow the various public and private insurance policies in place to play out without intervention, but this would have likely resulted in protracted individual settlements and major reconstruction and recovery delays (Rogers et al., 2014: 504). The land zoning process improved the pace of housing recovery activities, reduced the government's financial exposure to further earthquakes, and reduced the exposure of residents to ongoing disruption and threats to safety (especially in the Port Hills where there was significant risk from rock falls and cliff collapse).

Flood Zoning

The last major zoning-related decision addressed increased flooding vulnerability that emerged after the earthquakes. In March 2014, the need to seriously reconsider flood zoning in Christchurch became apparent as areas of the city, which had already been repeatedly battered by the earthquakes, experienced some of the worst flooding in decades. These floods were the combined result of heavy rainfall, seasonal high tides, and changes to the rivers and land caused by the earthquakes (Allen et al., 2014). The EQC covers increased flooding vulnerability if it is the result of land damage. So, after the earthquakes they led a process in Canterbury to gather updated topographical information; commenced a program of river, overland flow, and inundation modeling; and along with district councils initiated plans to update the boundaries of the Flood Management Areas (EQC, 2014). The Flood Management Areas are used to inform resource management and building consent, and the changed boundaries will cause tightened land-use controls for affected areas.

Although not its original intention, going forward, the residential red zone may be a key asset for regional flood hazard mitigation. The land in the residential red zone, much of which is near the Avon and Kaiapoi Rivers in Christchurch and the Waimakariri District respectively, may be used to increase water conveyance and detention capacity (Allen et al., 2014).

Lessons for Land Zoning

The three zoning processes discussed here evolved throughout the series of seismic events beginning in 2010 and processes were adapted and improved for the much larger February 2011 event. However, prior planning, systems, and lessons derived from international precedents were able to guide the post-earthquake zoning process. For example, the technical categories for the green zoning were, in part, informed by the concept of Zones of Required Investigation specified in the California Seismic Hazards Mapping Act 1990 (Rogers et al., 2014). It is therefore appropriate that further lessons are drawn from the zoning experiences in Canterbury.

These lessons can be applied both before earthquakes to reduce hazard risk, and after to improve recovery processes. First pre-disaster hazard assessments in Canterbury made decision makers aware of potential risks and in some cases allowed them to reduce exposure to these risks before the disaster event. While it is beyond the realm of practicality and affordability for all property owners in hazard zones to sell their property to the Government, ground improvement can be incorporated into renovation and development activities to reduce vulnerability (Rogers et al., 2014). In Christchurch, structures on ground that had been remediated prior to the earthquakes performed better than those on nearby 'untreated' ground (Wotherspoon et al., 2014). Additionally, incorporating hazard information into the requirements for foundations of newly built homes can increase their resilience.

Second, there are political, social, and economic costs and benefits of incorporating mitigation and vulnerability reduction into post-earthquake planning. The Christchurch CBD Red Zone, for example, slowed immediate recovery for those who would have had the means and motivation to pursue reconstruction activities quickly following the earthquakes. The zoning process has, however, facilitated more comprehensive planning in the CBD and coordinated vulnerability reduction in heavily damaged residential areas. Cordoning of commercial areas following large scale disasters, including the 2009 L'Aquila earthquake and the much more comprehensive example in Christchurch, may be considered a viable strategy for other places with a high concentration of vulnerable building stock (for example, the downtown area of many North American cities) (Chang et al., 2014). Zoning residential land and redrawing the boundaries of flood management areas not only reduced private and public exposure to future disruptions, but also provided markets and residents with an improved sense of certainty and security as they considered their options for recovery.

Enhancement of Lifeline Infrastructure Resilience

Lifelines infrastructure systems, including energy, water, telecommunications, roads, hospital systems, and ports, are crucial to the routine functioning of society and play an essential role in emergency relief, reconstruction and recovery of urban areas after an earthquake. Prior to the Canterbury earthquakes, there was already a strong "lifelines culture" in New Zealand. Policy and practice generally reflected an awareness of the essential role played by lifelines in the functioning of urban systems, and the necessity of enhancing their capability to continue functioning well during business-as-usual and after crises events.

The "lifelines culture" has been promoted in New Zealand by different local and national groups and government agencies, including among others robust assets, effective coordination at national and local levels, and risk-awareness for end-users. The New Zealand Lifelines Committee, for example, established in 1999, focuses on "enhancing the connectivity of lifeline utility organizations across agency and sector boundaries in order to improve infrastructure resilience" (NZLC, 2014). Similarly, the Civil Defense and Emergency Management Act 2002 mandates that lifeline utilities in New Zealand are "able to function to the fullest possible extent, even though this may be at a reduced level, during and after an emergency" (CDEM, 2002).

These mandates are carried out by infrastructure providers who are supported by the 17 voluntary Regional Lifelines Groups whose role is to identify infrastructure vulnerabilities and enhance resilience to regional scale emergencies. The EQC supports these Regional Lifelines Groups under the principle of "science to practice," promoting the transfer of knowledge and technology from basic research to professional practice across institutions, disciplines and sectors (EQC, 2012).

Lifeline Performance and Policy

The February 2011 earthquake in Canterbury created the largest lifeline disruption in a New Zealand city since the 1931 Hawke's Bay earthquake devastated Napier and Hastings. Road networks throughout Christchurch were extensively damaged by liquefaction-induced settlement. Local roads in the eastern suburbs of the city were the most affected, with 83 sections of 57 roads closed. Five of the 6 bridges crossing of the lower Avon River, which runs through Christchurch, were closed and many urban bridges required weight restrictions. The power, water, and sewage distribution networks were also badly affected. Substantial ground deformation induced by the earthquake caused multiple faults in the underground electricity distribution networks (50 percent and 14 percent of 66 kV and 11 kV cables respectively were damaged), inducing major power outages. About 30 percent of water and sewage pipes suffered extensive damage inducing severe and prolonged disruption to these services (Giovinazzi et al., 2011).

Approximately, 50 percent of the city was without water for the first days following the earthquake. The service was reestablished to over 95 percent of occupied units one month after; however, a boil order remained in-place for over six weeks for most of the city, because there was a risk that wastewater from the damaged sewage system had contaminated drinking water. The city also issued water conservation orders due to the damage to key water reservoirs and the loss of many groundwater pumping wells.

Sewage service was reestablished to 60 percent of occupied units one month after the earthquake. Thousands of remaining households relied on chemical and portable toilets distributed by the city council, until the service was gradually reinstated to all the customers six months after the earthquake. Due to severe damage at the main wastewater treatment plant for Christchurch, raw sewage was disposed in the rivers and estuaries for months following the earthquakes (Eidinger, Tang, and Rourke, 2010a).

Despite the catastrophic damage and disruption outlined above, the lifelines culture and programs for enhancing the resilience of the physical and organizational infrastructure systems established prior to the earthquakes helped mitigate the earthquake-induced physical and functional impacts and improved the effectiveness of recovery management. At least five key factors contributed to this resilience. First, new asset components were robust and resisted damage. Second, lifelines organizations had engaged in interdependency planning to understand how outages in other lifeline systems may affect them and created systems to continue functioning. Third, many lifeline providers had mutual assistance agreements. Fourth, lifeline providers tended to have extensive insurance cover. Finally, there was an overarching culture of readiness in lifeline organizations supported by highly trained and adaptable human resources.

Pre-earthquake seismic improvements to bridges on two major highways (Highways 73 and 74), for example, proved successful in resisting substantial loads and keeping the highways in operation post-earthquake. Similarly, in the mid-2000s the Canterbury-based power distribution company Orion, spent $6 million upgrading substation buildings (Orion, 2009). Despite the ground motions exceeding the design codes of the seismic strengthening program, only 1 of the 314 upgraded buildings failed. This seismic upgrade program was undoubtedly cost-effective, saving an estimated $30–50 million in potential damage and disruption following the earthquakes (Fenwick et al., 2011).

Organizational resilience also mitigated potential disruption following the earthquakes. All the lifelines utility organizations had mutual aid agreements and contingency measures in place that helped them to guarantee the prompt availability of materials and technical experts required for the repair operations. Most utility organizations also had back-up resources available that helped them to cope with the reduced functionality of other networks; the availability of back-up generators kept many of the systems functioning in the event of the many earthquake-induced power outages.

Several areas of underperformance and failure, however, moderated these successes. Here we discuss three notable issues: 1) aged asset components were vulnerable to disruption; 2) there were inadequate planning and tools to support post-disaster traffic management; and 3) there was no established process for infrastructure asset managers and technicians to access cordoned areas. Adaptive solutions to these problems emerged through the post-disaster process.

First, aged infrastructure is an important management consideration. While authorities were in many cases aware of system vulnerabilities it was often deemed too expensive to replace or upgrade infrastructure as part of risk mitigation programs. After the earthquakes, weak buried pipes and cables played a major role in the failures within the water, wastewater and power systems.

Next, traffic management was an unexpected and significant issue in the earthquake aftermath. The temporary traffic management of the city and highway network faced severe challenges to adapt to the damaged network and to the reorganization of the city, as businesses and residents relocated following the closure, demolition, and rebuilding of the CBD. This led to sustained and persistent traffic delays for residents trying to navigate the city and exacerbated logistical inefficiencies for businesses trying to operate after the earthquakes.

Finally, erecting large urban cordons was not established practice in New Zealand prior to the earthquakes, and therefore there were no procedures for managing access to areas within the cordon. This created delays, miscommunication, and frustration for infrastructure providers that needed to access their equipment behind the barriers. Utility operators could acquire a police escort upon request, but this process was sporadic and inconsistent.

Adaptive Resilience in the Infrastructure Response

One of the most significant adaptive responses that emerged during the earthquake recovery process was the creation of the Stronger Christchurch Infrastructure

Rebuild Team (SCIRT), the organizational alliance established to address the infrastructure rebuild in Christchurch. Together CERA and the initial Infrastructure Group formed within CERA, the Government, CCC, and NZTA selected an alliance-contracting model, embodied by SCIRT, through which a large-scale infrastructure rebuild could be undertaken.

Alliance contracting is a recognized way of achieving the best results from infrastructure programs in situations where project scope is uncertain, risks cannot be clearly defined and speed of repair is critical, as is almost always the case in a post-disaster context (Rotimi et al., 2009). Thus, SCIRT was formed and designated as the delivery team for managing the rebuilding activities for around 85 percent of the damaged horizontal infrastructure (including roads, fresh water, wastewater and stormwater networks) in Christchurch.

SCIRT set about achieving their vision statement of: "Creating resilient infrastructure that gives people security and confidence in the future of Christchurch" by enhancing the resilience of infrastructure systems in Christchurch (SCIRT, 2014b). It took several approaches for this, including making the asset components more robust, making them less likely to sustain damage in the event of future earthquakes, and making them faster to repair and re-establish services when subjected to damage. Infrastructure standards and specifications have been reviewed to ensure that they are fit for the new seismically active environment and represent value for money.

It is important to remember, however, that SCIRT was still operating with a limited scope and budget, and therefore not all desired resilience measure could be pursued. With their limited scope and resources in mind, SCIRT developed a decision support tool to decide how the infrastructure rollout could be prioritized to ensure a speedy and effective urban recovery (i.e. which infrastructure to be repaired, where, and when). Priority was given to the best approach from an engineering perspective, assessing need based on the physical condition of the asset, criticality within the system, the level of service, and the cost to maintain the service post-earthquake. However, SCIRT also considered community needs and services important to the local economy.

SCIRT's considerable achievements to date in embedding resilience within a post-disaster reconstruction program should be regarded as best practice. It represents best practice, not just for the pressure-cooker environment of post-disaster reconstruction, but also in ordinary business-as-usual environments. SCIRT's many innovations, from "competitive-collaborative" rebuild implementation, through to product and process technology developments, to staff secondment and HR management protocols are being captured, documented, and disseminated as part of the "SCIRT Learning Legacy project" (SCIRT, 2014a). This Learning Legacy project aims to ensure that the lessons from SCIRT can be transferred to others, both inside and outside of New Zealand.

Lessons for Lifelines Infrastructure Resilience

Some of the practical adaptive strategies developed following the earthquakes, such as post-disaster traffic management, and the lessons being captured in the SCIRT Learning Legacy project are actually just a small part of the learning that the Christchurch experience offers. The resilience mindset and work-streams geared toward infrastructure resilience enhancement created a snowball effect across all the infrastructure providers and agencies in New Zealand after the Canterbury earthquake sequence, and should be propagated worldwide.

Using the lessons from Christchurch to guide the resilience of New Zealand's infrastructure network, the National Infrastructure Unit (NIU), located within the NZ Treasury, is encouraging improved design standards, reviewing organizational culture to improve performance in emergencies, and identifying ways to quickly return services to full operational capacity. A number of other agencies in New Zealand are planning to implement the lessons from Christchurch and ensure that resilience is a key consideration when planning and operating infrastructure. For example, the NZTA have established a joint resilience operational policy project with New Zealand's rail and power infrastructure service providers, KiwiRail and Transpower (NIU, 2013).

The Canterbury earthquake revealed significant gaps in knowledge in the areas of:

- Pre-disaster mitigation (how to advance a justifiable and sustainable business case for infrastructure robustness and resilience pre-disaster as part of business-as-usual maintenance programs);
- Emergency response and short-term post-disaster recovery (the need to identify and assess the most efficient temporary solutions to reinstate infrastructure service post-emergency);
- Long-term post-disaster reconstruction (the need to identify and assess the most resilient reconstruction strategies and how to justify and fund the increased costs for them) (Liu et al., 2013; Orion, 2014).

As a result the earthquakes catalyzed considerable research and collaborative efforts to understand and enhance infrastructure resilience. The University of Canterbury in Christchurch, hosted an event called National Lifelines Week in 2011 and 2013, which aimed to bridge the identified knowledge gaps and to encourage a constant and proactive collaboration between researchers, stakeholders and utility managers (Brook, 2013). Another example of a significant collaboration emerging post-earthquake between researchers, stakeholders and infrastructure providers is the multi-disciplinary and multi-agency research project Economics of Resilient Infrastructure (ERI) funded by the New Zealand Government. The ERI project is ultimately geared toward developing a tool for a high-resolution assessment across space and through time of the economic consequences of infrastructure failure, business response and recovery options. The infrastructure response in

Christchurch following the earthquakes provides an excellent example of blending practical physical responses with organizational innovation to enhance resilience. It also offers several excellent examples of how disasters can be used as critical learning opportunities.

Policy Changes in the Construction Sector

In addition to the massive reconstruction effort required to restore and replace damaged lifeline infrastructure, Christchurch also requires a massive increase in construction activity to repair or replace hundreds of thousands of residential and commercial buildings. After any disaster, there is a need for the construction sector to respond quickly, using, or building on, existing capacities and arrangements. Three types of policy changes have been introduced in the construction sector in an attempt to expedite post-earthquake reconstruction and redevelopment, and to reduce future earthquake risks in Christchurch. These are: 1) new standards for house repairs and reconstruction following the Canterbury earthquakes; 2) changes to the building consent process; and 3) new regulation for upgrading existing earthquake-prone buildings. While the first two only affected construction processes in Canterbury, the third applies nationwide. By examining these policy changes, this section reveals the logic behind the amendments created to manage the building systems and performance of the New Zealand construction industry. It discusses what these regulatory changes offer that were not previously available and what lessons can be learned to help cope with future disasters.

Standards for House Repairs and Reconstruction

Changing building regulations or standards following a large disaster is a common policy response for reducing risks in future events (Wilkinson and Chang, 2011). In New Zealand, lessons from previous disasters had already been incorporated into mandatory codes. In the Building Code of New Zealand, design standards for buildings to reduce vulnerability to earthquake damage were introduced in 1935 following the 1931 Napier earthquake. Significant developments in design approaches have created a building code in New Zealand that is more stringent than many other countries. Some of the actions implemented following the Canterbury earthquake sequence reflect this readiness to incorporate new knowledge into building standards.

As discussed earlier, widespread liquefaction and lateral spreading following the Canterbury earthquakes meant that a large proportion of damaged properties needed new foundation designs to resist future land damage. For this purpose, housing decisions in the green zone subdivided land into three technical categories for liquefaction (TC1, TC2, and TC3 described in Table 6.2). The Department of Building and Housing[4] defined the three categories to describe how the land is expected to perform in future earthquakes, and also describe the foundation systems most likely to be required in the respective areas. In December 2010, the Department of Building and Housing published technical guidance on the repair and reconstruction of houses damaged in the September 2010 earthquake. This guidance was updated in November 2011 to include new land zoning decisions, and again updated in January 2013 to reflect new scientific and geotechnical information and knowledge about the impact of earthquakes and the effects of liquefaction on residential dwellings. In particular, the latest guidelines provide foundation solutions for earthquake-damaged houses in the land of TC3 (that is, land requiring the highest standard of foundation design).

The Engineering Advisory Group played a critical role in developing a consistent technical approach to housing repair and reconstruction. It was important that designers, insurers, and councils could all consistently interpret and apply these technical approaches. The advisory group, including structural and geotechnical experts from the public and private sectors, was initially established by EQC after the Darfield earthquake to investigate how residential structures responded to liquefaction effects. It was subsequently engaged by the Department of Building and Housing in developing its series of technical guidelines alluded to earlier, and acted as a committee to offer technical advice.

Changes to Building Consenting Process

Prior to the February 2011 earthquake, the building consent processes in New Zealand under the Building Act, was identified as a potential bottleneck to post-disaster reconstruction (Rotimi, 2010). This was, indeed, the case following the Canterbury earthquakes. The sheer volume of building consents required for the Christchurch rebuild overwhelmed the capacity of the Building Consent Authorities and led to prolonged delays. This made improving the consenting system a priority in post-earthquake Canterbury.

Although some efforts were made by the Christchurch City Council to streamline the earthquake recovery building work, criticism continued over the Council's inability to meet statutory timeframes for processing building consents. In a survey of construction organizations between November 2011 and January 2012, the time needed for getting a building consent was considered by the majority of respondents as the top issue affecting the recovery process (Chang and Wilkinson, 2012).

Reflecting the frustration experienced by homeowners and recovery agencies over the delays in consent processing and serious quality and management issues the International Accreditation New Zealand revoked Christchurch City Council's

4 This is now known as the Building and Housing Group of the Ministry of Business, Innovation and Employment.

capacity as a building consents authority, in July 2013. A Building Consents Action Plan was prepared by the Council to assess and improve the building consent systems and processes. Further in January 2014, the Christchurch City Council implemented a new management and team structure—Building Control and City Rebuild, which created a more streamlined process for commercial building consents.

New Regulations for Upgrading Existing Earthquake-Prone Buildings

The effects of the Canterbury earthquakes on buildings and subsequent human losses led to public scrutiny of the adequacy of policy settings and regulations for rectifying earthquake-prone buildings, and the effectiveness of their implementation and administration. During the earthquakes, most buildings in Christchurch performed in line with code expectations; however, there were still health and safety risks, especially arising from unreinforced masonry (URM) buildings (Ingham and Griffith, 2011). On the technical side, the unnecessary complexity of the building code was shown to compromise implementation and enforcement (Shelton, 2014).

In response to the Canterbury Earthquakes Royal Commission's recommendations, a comprehensive review of the system for managing earthquake-prone buildings was initiated in 2012. This review concluded that the system that was in place at the time of the earthquakes allowed buildings to remain that imposed an unacceptable level of risk and, therefore, did not adequately protect people from serious harm in moderate earthquakes (MBIE, 2013). To improve this system, the Government introduced the Building (Earthquake-prone Buildings) Amendment Bill in August 2013. The bill does not change the threshold level defined for an earthquake-prone building. It does, however, put the spotlight on the performance of buildings in earthquakes (Seville, Brunsdon, and Hare, 2014). The key decisions in the Bill stipulate that all commercial and large residential buildings will be assessed against potential earthquake damage within five years of this legislation taking effect, and all earthquake-prone buildings will have to be strengthened, or demolished, within 20 years of the legislation taking effect.

Lessons for the Construction Industry

The volume of repair and reconstruction activity required in Christchurch after the earthquakes placed considerable pressure on the damage assessment, engineering design, construction and consenting capacities available. In a paper on natural disaster legislative challenges written prior to the earthquakes, Rotimi et al. (2009) argued that in the New Zealand legislative context, the rigid policy prescriptions found in the conventional regulatory processes would not be helpful for disaster reconstruction. It is clear from the experience in Christchurch that the government has indeed gone beyond conventional building regulations after the earthquakes. Flexibility, the ability to accommodate change and the capacity and capability of technical experts are critical drivers behind those changes.

The organizations and institutions such as Engineering Advisory Group and Christchurch City Council's Building Control and City Rebuild Group were created to specifically limit risks and maintain operational focus after the earthquakes, and are bright spots for the future. They were successful because they had access to technology as well as a pool of experts and they had the flexibility to make technical decisions to achieve the safety goals envisioned in a post-disaster environment.

However, it needs to be recognized that for the regulations addressing earthquake-prone buildings to be successfully implemented, they may need to be accompanied by financial incentives for the building owner, whether in the form of tax credits, reduced insurance premiums, low-interest loans or exemptions from planning controls. The success of these regulations also depends equally on the development in society of a 'safety culture' in which both residents and building owners understand the risks they face and are prepared to participate in their management (Spence, 2004).

Policy Changes in the Insurance Sector

Much of the construction that needs to occur in post-earthquake Christchurch will be financed by insurance. Globally, New Zealand has a particularly high insurance penetration: in 2012 it ranked second in the world for non-life insurance (as a percentage of GDP) (CEBR, 2012). New Zealand is also ranked seventh in the world for commercial insurance cover (CEBR, 2012). One of the reasons for the high residential insurance penetration rates is the existence of the Earthquake Commission (EQC). EQC covers costs for any land damage[5] and damage to residential properties caused by geological events, up to $100,000 and $20,000 for contents. Private insurers then cover any losses over this amount, up to the insured value of the property.

The strong insurance base meant that the majority of the direct losses resulting from the earthquakes were covered by insurance: insured losses have been estimated at approximately 80 percent of the economic losses (Swiss Re, 2012). While high rates of residential and commercial insurance can been seen as an inherently desirable feature in a disaster-stricken area, it is important to reflect on how insurance shaped Canterbury's recovery and what lessons can be learned for the future.

5 EQC is one of the only insurers anywhere in the world to insure residential *land* for natural disaster damage as opposed to just homes and contents (EQC, 2013).

Impact of Insurance on Recovery

Insurance has provided much of the financial capital needed to drive the recovery of both businesses and residents in Canterbury. And because of this reliance on insurance financing, insurance processes and decisions have noticeably influenced the overall timing, cost, quality, and nature of the earthquake recovery. There are five issues in particular that have challenged the progress of insurance settlements and therefore slowed the pace of recovery.

First, the timeliness of recovery activities was affected by insurers struggling to deal with the sheer number of insurance claims and events. For EQC and many private insurers, each large earthquake had to be treated as a separate insurance event and new claims had to be lodged reporting any additional damage. EQC had anticipated 150,000 claims from an urban earthquake like this. More than 700,000 claims were received over the duration of the 2010/2011 earthquake sequence. To complicate this further, it was necessary for insurers to distinguish exactly what damage could be attributed to each specific 'insurance event' so that: 1) losses could be apportioned between EQC, private insurers and reinsurers; 2) insurers could determine how many deductibles to charge clients; and 3) how business interruption cover should be applied.

Finding and managing qualified personnel to assess and verify all these claims was an extreme challenge for affected insurers (Brown, Seville, and Vargo, 2013). EQC staffing levels alone jumped from 20 to 2,200 at its peak (King et al., 2014). As well as challenges in human resources, many insurers found their claims information management systems could not cope with the large number of multiple claims on properties. All these factors contributed to delays in insurance settlements and, therefore delays in subsequent recovery activities.

Second, insurance policy wording in some cases lacked clarity and invited inconsistent interpretation, which led to additional delays in insurance settlements. Terms such as "repair as new," for example, have been challenged by claimants as to whether it refers to a functional repair or total replacement of the damaged building or part of the building. This lack of clarity also led to increased payouts and numbers of building demolitions, with some claimants leveraging underspecified policy wording to gain higher value settlements. In other cases, it has meant that buildings have been demolished, not because they are dangerous but because they are "uneconomic to repair," in insurance terms (Muir-Wood, 2012). The high rate of demolitions has had complex consequences for Christchurch. It has created a significant amount of additional construction waste to manage, increased demand on building resources for the rebuild, and contributed to extended recovery timeframes. Conversely, this means that Christchurch is being rejuvenated with a new, better planned, more earthquake resistant building stock that will have benefits in the longer term.

Both insurers and policy holders have sought declaratory judgments to confirm how to interpret policies correctly (*Earthquake Commission v The Insurance Council of New Zealand Inc, Vero Insurance New Zealand Ltd, IAG New Zealand*

Ltd. Tower Insurance Ltd v Earthquake Commission., 2011; *Sunny Alison June Morley and New Zealand Trustee Services Limited and Mt Vision Limited vs Earthquake Commission*, 2013), which again contributed to delays in processing insurance settlements.

Third, there were several complex technical challenges during the recovery that tested insurers and their policies: some areas of Christchurch city were abandoned due to land damage and rockfall; new regulations were brought in to improve the quality of foundations in areas subject to liquefaction; the Building Code was changed to reflect increased seismic activity; and the City Council required that all buildings being repaired be brought up to 67 percent of the New Building Standard capacity (King et al., 2014). Since many insurance policies were full replacement, one of the biggest challenges was determining which changes in regulations and standards insurers were required to adhere to and which changes were the financial responsibilities of the property owner. Again individuals, organizations, and insurers sought declaratory judgments to determine insurer liability (*Helen Kraal and Bruce Roberston Irvine vs The Earthquake Commission and ALLIANZ New Zealand Ltd*, 2014; *Insurance Council of New Zealand Incorporated vs Christchurch City Council and University of Canterbury and Body Corporate 423446 (Oxford Body Corporate)*, 2013). For example, in the Insurance Council vs Christchurch City Council et al., the Insurance Council argued (and were proved correct though this is still being contested in the courts) that the Christchurch City Council could not legally compel building owners to strengthen their buildings to any more than 34 percent of the New Building Standard capacity during repair work. Therefore, insurers were not liable to pay for earthquake strengthening works. Once the courts determined liability, insurers and property owners had to compare relative feasibility of repair or replacement strategies—which often required several iterations of design options and costs for commercial properties. These issues complicated and delayed claims settlement processes and subsequent repair works

Fourth, doubt around future insurability in Christchurch generated a level of uncertainty in the recovery process and decision making. Insurers were reluctant to insure new risks following the earthquakes: after every significant earthquake, a 21-day moratorium was placed on new policies. This made it difficult for repair and rebuild works to be carried out, and for investors to make informed decisions about re-investing in the city.

Fifth, most insurers elected to manage their clients' earthquake repairs themselves, using appointed contractors as opposed to providing cash payment. Insurers pursued this option in an effort to control building costs and to, therefore, reduce losses for insurers. Increased building costs, or demand surge, is often present in post-disaster environments where demand for certain goods and services is high and insurers wanted to minimize these cost increases through a robustly managed repair process. As a result of this process, residents and businesses were sometimes restricted to a repair program governed by their insurers. In some cases this also meant that building owners had less scope to innovate and build their properties back better (than what they were insured for) as insurers were unwilling to extend the scope of their contractor's responsibilities.

The delays and uncertainty in the recovery process has had a noticeable effect on both homeowners' and businesses' recovery decisions and actions. Some home owners have been waiting over three years to settle their claims, some of whom are still living in badly damaged properties. This has led to increased levels of stress in the community (King et al., 2014). For businesses, delays in settling material damage claims has often meant that business interruption periods have finished prior to disruptions caused through repair and rebuild works. Many businesses have had to rely on organizational cash flow, rather than insurance payments, to fund their recovery (Stevenson et al., 2011). As noted above, delayed insurance payments and confirmation of future insurance coverage have also impacted some developers' investment decisions in Christchurch. As well as delays, the high level of insurance, lack of clarity in some policies and the degree of repair management that insurers assumed, has contributed to a higher level of property replacement (demolition and rebuild) and reduced scope for business innovation. Despite all this, it must be noted that surveys have shown that the majority of businesses were satisfied with their insurers and insurance payments (Kachali, 2013; Stevenson et al., 2011).

Lessons about Post-Disaster Insurance

It is without doubt that the high level of insurance in New Zealand has provided a solid financial basis for the recovery and new opportunities from which to grow. However, there are many lessons that can be learned about how to manage such an immense insurance event and avoid the delays and other negative outcomes discussed above.

A number of valuable lessons have already been heeded by insurers operating in New Zealand and subsequent changes have been made to policies and pricings. In the residential sector, the major change in insurance is a move from full replacement insurance to 'sum insured' policies. Insurers want to cap their liability and avoid the escalation of losses that occurred in Christchurch. For both residential and commercial properties, there is less availability of insurance for high-risk properties (for example unreinforced masonry or houses on unsuitable foundations). There is a general increase in insurance premiums and deductibles, particularly for commercial properties, and some insurers have changed how deductibles are calculated—moving from a percent of claim value, to a percentage of insured value (ICNZ, 2012). Many insurers have also revisited their standard wording around reinstatement of cover to make it clear whether insurance is automatically reinstated after any event or if they limit their liability to the sum insured for any given insurance period (typically a year).

In addition to changes already made by insurers to their policies, there are other lessons that should be considered, in particular how to reduce delays in claims settlement and the effects of those delays. Potential approaches include: clarifying relationships and claim assessment procedures between dual insurers (in this case EQC and private insurers); providing streamlined and prioritized claims settlements for vulnerable individuals and businesses; adoption of standard policy wordings and interpretations; simplifying settlement options; providing interim payments for insurance where possible (particularly business interruption); and allowing deferment of business interruption payments (Brown et al., 2013).

Strategies Adopted to Reduce the Economic Impact of the Earthquakes

Although insurance is financing much of the reconstruction of damaged buildings, sustained economic viability in a revived city requires a much more comprehensive approach. In 2003, the Ministry for Civil Defence & Emergency Management identified economic recovery as an integral component of their holistic framework for disaster recovery. The 'Economic Environment' in this framework is composed of four distinct elements: individuals, firms, infrastructure, and government (Norman, 2006). Though specific economic support programs were not outlined in the 2003 framework for holistic disaster recovery, many of priorities identified for each element were addressed in the aftermath of the Canterbury earthquakes.

The earthquake had temporary negative impacts on employment and consumer spending in Canterbury. A Department of Labour survey in 2011 indicated that up to 60 percent of workplaces in the professional, scientific and technical sector had to relocate all or part of their operations. Around 40 percent of businesses reported lower revenues and decreased customer numbers (Chang-Richards et al., 2013; DOL, 2011). At the time of the 2013 census more businesses were still migrating out of Canterbury than into the region. However, economic activity in Christchurch was around 95 percent of pre-earthquake levels within 18 months of the February 2011 earthquake (CERA, 2012a). By 2013 overall regional employment exceeded pre-earthquake levels (CERA, 2013).

The overall effect of the earthquakes on the regional economy were relatively subdued and there were few pronounced impacts on the aggregate New Zealand economy (Dalziel and Saunders, 2012; Doyle and Noy, 2013). Part of the reason for this was policies implemented by the Crown and the Reserve Bank that were not connected directly to the earthquakes, including increasing GST (Goods and Services Tax) in October 2010 (Doyle and Noy, 2013). The economic impact of the earthquakes was, however, also mitigated by direct policy interventions geared toward reducing business downtime and capital flight, avoiding economic stagnation, and promoting investment in an environment with new challenges and opportunities. In keeping with MCDEM's holistic recovery framework decision makers pursued policy initiatives that mitigated some of the earthquake's potential negative economic impacts. These policies were aimed at the micro-level, to individuals and firms, and on a meso- and macro-level, to support affected local governments and local urban economic ecologies.

Individuals and Firms

Four broad policy approaches helped individuals and firms manage and reduce the economic impact of the earthquakes. These interventions included insurance interventions, income assistance, tax relief, and guidance for firms navigating the recovery environment.

Insurance

As discussed in the previous section, EQC insurance was a significant source of financial capital for residents affected by the earthquakes. As a result of this program, the national government and private insurers will cover the brunt of the financial cost of residential repairs and reconstruction. Additionally, the national government through CERA worked directly with private insurers on issues regarding the availability of insurance and the timely resolution of claims for businesses (CERA, 2012a).

In some respects, however, the government may not have gone far enough to intervene in the settlement of insurance claims. The cost and availability of commercial insurance is still one of the most significant limiting factors in the progress of the rebuild. The New Zealand government and other areas at risk from major and repetitive disasters may need to consider subsidies or greater degrees of intervention to enhance private insurance coverage for commercial clients. There is some international precedent, demonstrating broader interventions in private insurance markets in the wake of a major disaster. Following the 9/11 terrorist attacks, for example, the US government passed the Terrorism Reinsurance Act (TRIA), a federal reinsurance support system that ensured that customers could still access affordable private insurance in a market with greatly heightened risk (AON, 2014). It was considered highly successful to stabilizing the commercial property insurance market after these attacks (Zandi et al., 2006).

Income Assistance

Soon after the September 2010 earthquake, decision makers at the national level developed policy to provide income assistance to individuals and businesses affected by the earthquakes. This policy addressed concerns that earthquake affected businesses would lay-off staff, and as a result a large number of people would either join the unemployment benefit or leave the region to find work. The Government through its national social assistance agency Work and Income New Zealand (WINZ) provided an Earthquake Employment Support package with two options available:

The Earthquake Support Subsidy (ESS) was eligible to small- to medium-sized enterprises affected by the disaster. They received a weekly payment to help pay wages for each full-time and part-time employee.[6] In most cases the firms receiving the ESS were viable, but experiencing relatively short-term closure, productivity, or cash flow problems. Earthquake Job Loss Cover (EJLC) was eligible to employees whose employer believed their business is no longer viable or who were unable to make contact with their employer following the earthquakes.[7]

These two programs had manifold benefits for businesses, employees, and the economy. The ESS temporarily relieved businesses of the pressure to make staff redundant. It also prevented some businesses from rushing to reopen and committing to suboptimal leases or other recovery decisions before they were ready. The ESS also allowed the government to provide financial assistance without disrupting businesses' relationships with their employees. For businesses the ESS was a safer and easier option than obtaining loans (e.g., the subsidized post-disaster Small Business Association loans available to disaster-affected businesses in the US) because there were no requirements to prove financial viability and they did not acquire additional debt. The EJLC, on the other hand, helped workers directly as they either transitioned to new jobs or connected with other welfare assistance

Both of these programs required a minimal burden of proof for eligibility. As a result, employees and employers accessed this support quickly, uptake rates were relatively high, and public opinion of the program was overwhelmingly positive. Despite the risk of distributing public money without a proper evaluation process, early appraisals indicate that the ESS and EJLC ultimately saved the government money due to the efficiency of its deployment and its timeliness in buoying disrupted small businesses (Fischer-Smith, 2013). While the EJLC was based on existing welfare payments, nothing like the ESS had been deployed in New Zealand prior to these earthquakes and it therefore represents a successful innovative response.

Tax Relief

In 2012 and 2013, late filing of taxes and payment non-compliance increased significantly as earthquake-affected businesses struggled to comply with their legal and financial obligations (IRD, 2014). New Zealand's national taxation body, the Inland Revenue Department (IRD) reacted to this issue by creating a small to medium sized business support package which included: waiving fees for filing taxes late; providing tools to help organizations self-manage better; and offering seminars, meetings, and expos for businesses. By implementing a responsive tax regime, the IRD intended to ease the recovery process for affected firms and ease regional economic recovery.

---

6 Employers received NZ$500 for each full-time employees and NZ$300 for each part-time employee.

7 Through the EJLC program full-time workers received NZ$400 net per week and part-time workers received NZ$240 net per week regardless of the pre-earthquake income.

Tax compliance rates improved through 2012. However, research conducted by Inland Revenue indicates that the removal of late fees caused some businesses to feel a reduced sense of urgency to become tax compliant. As a result tax compliance rates in Canterbury, remained worse than they had been prior to the earthquakes. Thus, in 2013 Inland Revenue reverted back to business-as-usual to not only help small and medium sized businesses return to better compliance levels, but also "help decrease hidden economy activities" (IRD, 2014, p. 2).

Guidance for Businesses

In March 2011, the Canterbury Employers Chamber of Commerce and the Canterbury Development Corporation[8] combined to form Recover Canterbury. Recovery Canterbury was supported by several government agencies, and was intended to provide small to medium sized businesses with guidance and support following the earthquakes. They set up mobile business recovery centers, hired business recovery coordinators to advise businesses, ran workshops and networking events, and administered the Canterbury Business Recovery Trust Fund which provided grants directly to organizations in need following the earthquakes.

Financial assistance provided through Recover Canterbury aligned with the Government's principles for business support, which emphasized targeted projects with specific planned outcomes (Hatton, Seville, and Vargo, 2012). About 7,000 businesses received some form of support. Of those, about 400 businesses received some portion of the $6.1 million of funding distributed through Recovery Canterbury (Recover Canterbury, 2012). The Canterbury Development Corporation assessed the economic impact of Recover Canterbury in 2012, estimating that it saved over 600 jobs and kept an estimated $39 million in the Canterbury economy (Recover Canterbury, 2012). It is important to note that the Canterbury Development Corporation was directly involved in Recover Canterbury, and is therefore not an independent assessor. However, the Canterbury Development Corporation estimate reflects a broader sense from business owners with whom the authors have interacted, that Recover Canterbury provided an invaluable source of support to affected businesses in Canterbury. Yet, opinions of the program were not as universally positive as they were for the ESS, in part, because Recovery Canterbury required significant vetting and proof of financial viability before distributing vouchers and grants. Recover Canterbury did, on other hand, provide a much broader scope of services for businesses and was available for a longer period of time than the ESS.

---

8 The Canterbury Employers Chamber of Commerce is a not-for-profit organization and business support agency for the Canterbury region. The Canterbury Development Corporation is the economic development agency for the Christchurch City Council.

Supporting the Labor Market

The labor market in Christchurch faced challenges as some industry sectors, including retail and tourism, contracted, while the construction industry experienced significant expansion. With minimal skill transferability between contracting and expanding sectors and the anticipated need for between 20,000 and 30,000 extra construction-related workers (CERA, 2012b), the government implemented programs geared specifically toward fulfilling the construction labor demand (Chang-Richards et al., 2013).

These plans included setting up the Canterbury Employment and Skills Board to develop a Labor Market Plan and conduct workforce assessments. The 2011 Central Government budget also included the "Skills for Canterbury" package which: (a) expanded funding for construction-related training at NZ Polytechnic Institutions and other industry training organizations in Canterbury; (b) increased funding for training and work placement in Canterbury; (c) increased funding for Industry Partnership placements to help unemployed people take up reconstruction related jobs; and (d) created a special immigration skills shortage list that made it easier for employers to hire people from overseas to meet immediate labor needs (CERA, 2012b).

It can be difficult to evaluate the overall effectiveness of these policies. For example, the tertiary education initiatives for engineering degrees and other skilled labor are part of a longer-term expansion and development process. The benefits of this kind of policy necessarily have a delay of several years before participants can complete the necessary training and assist the rebuild (Chang-Richards et al., 2013).

Chang-Richards et al. (2013) suggest that the measure of labor market outcomes, following a disaster, should be evaluated on the numbers of unemployed or the operational status of businesses, as well as on the answer to two questions: "1) How and where will those who are displaced from their jobs be replaced in the short-term? And 2) how and where will those most vulnerable (women and youth) find employment opportunities for longer-term benefit?" (p. 15). In Christchurch, those who lost their jobs in badly affected industries (e.g., retail and hospitality) were not the people who could fill vacancies in growth industries. Additionally, women and young people experienced relatively higher unemployment rates than other groups immediately following the earthquakes. Although individual income assistance programs like the EJLC provided some temporary relief, long-term labor policies following the Canterbury earthquakes did not specifically target programs toward these vulnerable groups.

Support for Local Governments

In addition to support for individuals and firms it was also critical to the long-term economic recovery of Canterbury that recovery policy worked at the meso- and macro-economic scales. Policies needed to consider the interactions and functions of the wider institutional and urban environment. An important component of this policy approach was helping local governments and local commercial areas remain financially solvent and able to resume and continue functioning.

Cost Share Agreement

In 2013 the national government and the Christchurch City Council signed a Cost Sharing Agreement. This agreement set out the proportion of funding the national government would provide for major central city projects and repairing and replacing horizontal infrastructure. This was done on a percentage basis, with the national government committing to a significant proportion of the cost for many of the major rebuilding initiatives, including for example covering 60 percent of costs on water infrastructure repairs and over 80 percent of all costs for road repairs (CCC, 2013). Overall the national government is contributing $2.9 billion to the rebuild and the Christchurch City Council is contributing $1.9 billion (CCC, 2013).

The Christchurch City Council described the Cost Sharing Agreement as one of the most important steps forward for earthquake recovery in the city. It provided the security necessary for the Council to proceed with planning for large anchor projects and infrastructure renewal.

Planning in the Christchurch CBD after the earthquakes focused on anchor projects and on fostering industry clusters of "precincts" to encourage medium and long-term return of businesses to the CBD. These precincts are intended to facilitate collaboration and economies of scale and investment. They also attempt to remedy past issues with urban sprawl and the oversupply of commercial space. The Christchurch CBD recovery plan, developed by Christchurch Central Development Unit,[9] includes several sector-specific precincts (e.g., retail, health, and performing arts), and "The Frame," a border of green space around the city designed to "contain the Core" of the city (CCDU, 2012, p. 31). Many of these precincts are anchored by a government building or hub with significant government funding. For example, the Christchurch Hospital is expected to receive an estimated $500 million for redevelopment, and will anchor the Health Precinct in the central city. Thus, without the strong financial and logistical support of the national Government, planning and recovery could not progress toward an improved layout and resilient future.

9 In April 2012, CERA established the Christchurch Central Development Unit (CCDU) which was tasked with laying out a plan for the CBD. Many raised concerns that the CCDU was not established collaboratively with the Christchurch City Council and in some ways trumped the effort put into the planning and consulting process the CCC had undertaken. Some Christchurch leaders, however, felt that the CCDU was necessary to "overcome the number of small land titles and the complexities of existing use-rights, as well as to create confidence in the CBD through the siting of major [anchor] projects" (Taylor et al., 2012: 14).

Despite this important cooperative financial agreement, progress on precincts and anchor projects in the Christchurch CBD is very limited as of 2014. This is, in part, due to protracted conflicts between commercial property owners and the Christchurch Central Development Unit during the land procurement process. Additionally, a perceived lack of alignment and poor communication between the Central Government and the Christchurch City Council may be having a negative effect on private investor confidence (NZCID, 2013). Thus, an important final component of progressing the CBD rebuild is building the confidence of private investors that will provide the momentum and long-term sustainability for these projects.

Support for Local Commercial Centers

On a smaller scale the national and local government supported initiatives to support and promote badly affected commercial centers in neighborhoods and suburbs affected by the earthquakes. Business associations and economic development agencies received funding from the central government to pursue local promotion campaigns that would draw people back into areas that had been damaged. In smaller suburban areas and town centers, business associations used government funding to, for example, run campaigns to encourage residents to support local businesses or attract visitors to the area.

Local organizations also worked with the city council to create spaces for social engagement where normal patterns of interaction had been disrupted by closed roads, damaged buildings, and cordons. A notable example is the ReStart Mall in the Christchurch CBD. Within a month of the February 2011 earthquakes, business advocates and members of the Central Christchurch Business Association started the Restart the Heart Trust.[10] The purpose of the trust was to establish a central retail hub in shipping containers populated by "iconic" CBD retail and hospitality businesses (Stevenson, 2014). The ReStart Mall opened in October 2011 in an area that was previously a central pedestrian mall. In its early stages it served as a major draw for people to maintain contact with the city center and a place for displaced businesses to operate in the central city, though interest waned with time.

Promotion campaigns and the ReStart Mall did not prove to be financially lucrative for many of the individual businesses involved; however, they did help maintain ongoing citizen engagement with these damaged commercial centers during the rebuild. Due to building damage, depopulation, and the cordons, the economic "ecology" (that is, the web of interactions among businesses and with their customers that create synergies and communities of support) in the Christchurch CBD and other local commercial center, was severely disrupted and

10 The project was funded largely though sponsorship from ASB Bank and the Christchurch Earthquake Appeal Trust, an independent charity responsible for allocating money raised by the Christchurch Earthquake Appeal.

in some places destroyed. Thus, the ReStart Mall was an important intervention in the CBD to help support ongoing interaction with the central city area.

Lessons for Economic Recovery

Many of the policies pursued to mitigate the negative economic effects of the earthquake and promote sustainable economic recovery were in keeping with MCDEM's 2003 framework for holistic disaster recovery. For example, the framework discusses maintaining individual livelihoods, including supporting employment security and payment of wages. Although there was no policy precedent, this need was met after the 2010/2011 earthquakes using the ESS and EJLC programs. The 2003 framework also prioritized a strong infrastructure restoration program as an essential component of economic recovery and avoiding cascading economic failures. Section 3 illustrates that infrastructure restoration and resilience building has been a core focus of the recovery.

It is important to recognize that disasters present businesses and individuals with a number of new complexities and challenges. Thus, one of the most significant lessons derived from Christchurch's experience is the importance of policy that eases access to resources and removes barriers to normal business functioning (while still maintaining the broader goals of socially and economically sustainable recovery). For example, the ESS was a success in large because it allowed businesses to access stopgap financial support with very minimal burden of proof. Businesses that had lost files, could not access their buildings, and were dealing with other earthquake stressors, may have fallen through the cracks if the ESS vetting processes had been more rigorous. Similarly, the IRD's decision to remove fees for late tax filing, stopped hundreds of businesses from accruing debt during the recovery process that may have led to long-term dysfunction and failure. Policy also eased access to labor by supporting flexible immigration processes to fill skill shortages. These policies all sought to support, rather than supplant normal business function.

Outcomes at the city level are more difficult to evaluate. Christchurch has yet to realize the benefits of the co-coordinated urban development process in the Central Business District. Time will tell whether the delays created by cordon and the centrally coordinated land procurement process will pay off. The process may still result in a better planned more vibrant city. It could also leave Christchurch with an anemic core if the economic and social activity that has been displaced to the city periphery does not return to the center. The processes unfolding in Christchurch will continue to produce important lessons for urban centers recovering from disasters.

Conclusion

In 2013, Christchurch became one of the inaugural cities to join the Rockefeller Foundation's 100 Resilient Cities Challenge. The Challenge is intended to develop a global network of cities focused on developing robust and innovative approaches to building urban systems that can deliver basic functions to every part of the population in all conditions. Involvement in the Challenge represents one part of a wider effort of Christchurch's community, economic, and policy leaders' efforts to incorporate vulnerability reduction and resilience enhancement into the city's reconstruction and recovery.

In this chapter, we have examined several main components of the urban system that were heavily impacted by the 2010/2011 Canterbury earthquake series: land, infrastructure, building construction, insurance, and the economy. In each section we provided a reality check, assessing the post-earthquake performance of each component compared to pre-earthquake planning and mitigation. We then explored the way systems were necessarily adapted and altered to accommodate new realities and information that emerged as a result of the disaster and throughout the early stages of recovery.

For each component of the urban system we presented lessons, derived from Christchurch's ongoing process of reconstruction and recovery, which can be used to support planning and recovering from urban disasters internationally. Experiences with land zoning after the earthquakes revealed the value of pre-disaster hazard assessments and land remediation. This process also highlighted the tradeoffs inherent in pursuing risk reduction post-disaster, and the difficulty of weighing short- and long-term costs. For example, in Canterbury the Government had to balance the short-term cost of the massive residential property buy-out against the long-term costs of rebuilding infrastructure in vulnerable areas and the risk of stagnated land-markets.

The lifeline infrastructure and building reconstruction have also produced lessons. Reconstruction in Christchurch is being progressed through collaborative and innovative processes. The response of public and private infrastructure and construction organizations has demonstrated the value of establishing positive cultures of resilience thinking and inter-organizational communication before a disaster. These cultures facilitate the flexibility that is necessary to accommodate change and overcome rigid and prescriptive practices established during business-as-usual. The reconstruction process has also provided several useful examples of systems for assessing knowledge gaps and capturing knowledge generation following a disaster.

Insurance is a dominant player in the story of Christchurch's rebuild. The high degree of coverage and the ongoing nature of the disaster have led to a large number of claims and resulting delays. These delays have been exacerbated by contestable policy wording. The insurance industry in New Zealand has already implemented lessons learned during this crisis to improve policy wording and reduce risk exposure in future events. This action by insurers, however, has shifted a greater proportion of future financial risk back onto home and business owners. This demonstrates the importance of pursuing mitigation and risk reduction, rather than relying exclusively on risk displacement. The efforts to reduce the negative economic impacts of the earthquakes in Christchurch reveal the benefits of interventions that enable rather than displace normal business activity. However, the efforts at centrally coordinated urban planning and economic renewal in the Christchurch CBD have, so far, proven problematic and it is unclear whether this approach will pay off in the long run.

From these various aspects of the Christchurch reconstruction and revival experience, we can draw four general insights that will help institutions and decision makers navigate the post-disaster urban environment. First, pre-disaster mitigation and preparedness are an essential part of business-as-usual practice. Hazard and risk assessments conducted prior to the Canterbury earthquakes were instrumental in facilitating risk reduction prior to the earthquakes. This work, however, was not always accounted for in cost-benefit assessments for infrastructure and building construction and maintenance. When decision makers accounted for hazard risks and built resilience into engineered and organizational systems before the earthquakes it minimized loss and those systems experienced more efficient returns to functionality.

Second, recovery effectiveness is enhanced if core urban systems can quickly resume functioning. This includes infrastructure and local government, but more importantly social and economic networks need to be supported and maintained if the urban system is to sustainably repair itself through continued growth and evolution.

Third, no amount of planning and mitigation will be adequate to address every situation following a major urban disaster. It is therefore essential that policy and organizational systems reduce rigidity and develop adaptive capacity prior to a crisis and create room for flexibility and innovation after a crisis.

Finally, auditing and monitoring decisions and actions and incorporating lessons learned needs to begin immediately following a disaster and needs to continue throughout the entire response and recovery process. The ongoing nature of the earthquake series in Christchurch demonstrated that it is never too soon to improve and prepare for the next disruption. Lessons incorporated into policy and organizational systems following the September 2010 earthquakes eased response and recovery efforts following the February 2011 earthquakes. In dynamic environments learning, communicating the lessons learned, and incorporating those lessons into the appropriate systems is an iterative and continuous process, and essential to successful and sustainable urban recovery.

Acknowledgments

The authors extend their sincere thanks to Canterbury Earthquake Geospatial Research Fellow, Dr Matthew Hughes for his help with the preparation of this chapter. Much of the knowledge shared in this chapter was generated through the extensive research conducted by the authors and others in the Resilient Organisations research program (www.resorgs.org.nz) in their relevant areas of expertise since September 2010. This work would not have been possible without the generous cooperation and input of hundreds of informants, research participants, and collaborators.

References

Allen, J., B. Bradley, A. Bradshaw, et al. (2011). *Geotechnical Reconnaissance of the 2011 Christchurch, New Zealand Earthquake* (No. GEER-027). http://www.geerassociation.org/GEER_PostEQReports/Christchurch_2011/Cover_Christchurch_2011.html.

Allen, J., C. Davis, S. Giovinazzi, and D. Hart (2014). *Geotechnical and Flooding Reconnaissance of the 2014 March Flood Event Post 2010–2011 Canterbury Earthquake Sequence, New Zealand* (No. GEER-035). http://www.geerassociation.org/GEER_PostEQ Reports/Christchurch_Flood_2014/FinalReport19June 2014.pdf.

AON (2014). Terrorism Risk Insurance Act: The Case to Extend. *AON Risk—Terrorism Risk Insurance Act (TRIA): Aon's Resonse to the U.S. Treasury and President's Working Group*. http://www.aon.com/risk-services/thought-leadership/reports-pubs_terrorism-risk-insurance-act.jsp.

Bascand, G. (2011, March 24). Gross Domestic Product: December 2010 quarter. *Statistics New Zealand*. http://www.stats.govt.nz/browse_ for_stats/economic_indicators/GDP/GrossDomesticProduct_HOTPDec10qtr.aspx.

Brook, K. (2013). National Lifeline Week at UC in December. *University of Canterbury Communications*. http://www.comsdev.Canterbury.ac.nz/rss/news/?feed=news&articleId=1068.

Brownlee, G. (2012, August 31). *State of Canterbury Earthquake Recovery*. http://beehive.govt.nz/speech/state-canterbury-earthquake-recovery.

CAENZ (1997). *Risks and Realities: A Multi-disciplinary Approach to the Vulnerability of Lifelines to Natural Hazards* (1st edition). Christchurch, NZ: Centre for Advanced Engineering New Zealand.

CCC (2011). *Christchurch Earthquake Response: Liquefaction Frequently Asked Questions*. Christchurch, NZ: Christcurch City Council, 2. http://canterbury earthquake.files.wordpress.com/2011/03/liquefaction-factsheet-20110316.pdf.

CCC (2013). *Cost Sharing Agreement*. Christchurch, NZ: Christchurch City Council. http://cera.govt.nz/sites/default/files/common/cera-crown-ccc-cost-sharing-agreement-2013-12-05.pdf.

CCDU (2012). *Christchurch Central Recovery Plan Te Mahere Maraka Ōtautahi*. Christchurch, NZ: Central City Development Unit.

CDEM (2002). *Civil Defence Emergency Management Act 2002*. Wellington, NZ: New Zealand: Ministry of Civil Defence and Emergency Management. http://www.legislation.govt.nz/act/public/2002/0033/latest/DLM149789.html.

CEBR (2012). *Lloyd's Global Underinsurance Report*. London: Centre for Economics and Business Research Ltd. http://www.lloyds.com/~/media/Files/NewsandInsight/360RiskInsight/Global_Underinsurance_Report_311012.pdf.

CERA (2012a). *Economic Recovery Programme for Greater Christchurch*. Christchurch, NZ: Canterbury Earthquake Recovery Authority. http://cera.govt.nz/recovery-strategy/economic/economic-recovery-programme.

CERA (2012b). *Economic Recovery Programme: Labor Market*. Christchurch, NZ: Canterbury Earthquake Recovery Authority. http://cera.govt.nz/recovery-strategy/economic/labour-market.

CERA (2012c). *Land Zones*. Christchurch, NZ: Canterbury Earthquake Recovery Authority: Property Information. http://cera.govt.nz/land-information/land-zones.

CERA (2012d). *Overview of TC1, TC2, and TC3 Technical Categories*. Christchurch, NZ: Canterbury Earthquake Recovery Authority: Property Information. http://cera.govt.nz/residential-green-zone-technical-categories/overview.

CERA (2013). *Canterbury Economic Indicators: August 2013*. Christchurch, NZ: Canterbury Earthquake Recovery Authority. http://cera.govt.nz/sites/default/files/common/canterbury-economic-indicators-quarterly-report-august-2013.pdf.

CERA (2014). *Residential Red Zone*. Christchurch, NZ: Canterbury Earthquake Recovery Authority. http://cera.govt.nz/residential-red-zone.

Chang, S.E., J.E. Taylor, K.J. Elwood, et al. (2014). Urban disaster recovery in Christchurch: The central business district cordon and other critical decisions. *Earthquake Spectra* 30(1): 513–532, doi:10.1193/022413EQS050M.

Chang, Y. and S. Wilkinson (2012). *Preliminary Results from the Resource Availability for Christchurch Earthquake Reconstruction (RACER) Survey*. Auckland, NZ: The University of Auckland. http://www.recres.org.nz/publications.

Chang-Richards, A.Y., E. Seville, S. Wilkinson, and B. Walker (2013). *Building Natural Disaster Response Capacity—Sound Workforce Strategies for Recovery and Reconstruction: Final Report*. Singapore: Asia-Pacific Economic Cooperation.

Dalziel, P. and Saunders, C. (2012). Regional development before and after an earthquake: The Canterbury New Zealand experience. *Australasian Journal of Regional Studies* 18(1): 100–127.

DOL (2011). *A Changing Landscape : The Impact of the Earthquakes on Christchurch Workplaces*. Wellington, NZ: Department of Labour. http://www.dol.govt.nz/publications/research/earthquakes-impact-in-christchurch-workplaces/earthquake-impact-in-christchruch.pdf.

Doyle, L. and I. Noy. (2013). *The Short-run Nationwide Macroeconomic Effects of the Canterbury Earthquakes* (SEF Working paper: 01/2013 March 2013). Wellington, NZ: University of Wellington: School of Economics and Finance.

ECan (2007). The Solid Facts on Christchurch Liquefaction. *Environment Canterbury, Regional Council.* http://ecan.govt.nz/publications/General/solid-facts-christchurch-liquefaction.pdf.

ECan (2014). Floodplain strategies and historic flood reports. *Environment Canterbury, Regional Council.* http://ecan.govt.nz/advice/emergencies-and-hazard/flooding/Pages/floodplain-strategies-reports.aspx.

Eidinger, B.J., A. Tang, and T.O. Rourke (2010). *Technical Council on Lifeline Earthquake Engineering (TCLEE) Report of the 4 September 2010 Mw 7.1 Canterbury (Darfield), New Zealand Earthquake*. Reston, VA: Technical Council on Lifeline Earthquake Engineering. http://www.asce.org/uploadedFiles/Technical_Groups_and_Institutes/TCLEE/TCLEE%20Preliminary%202010%20NZ%20EQ%20Report.pdf.

EQC (2012). Lifelines groups, in *Earthquake Commission: Projects, Programmes and Sponsorships*. Wellington, NK: Earthquake Commission. http://www.eqc.govt.nz/research/projects/lifelines-groups.

EQC (2013). *About EQC*. Wellington, NZ: Earthquake Commission. http://www.eqc.govt.nz/about-eqc.

EQC (2014). *Land with an increased risk of flooding*. Wellington, NZ: Earthquake Commission. http://www.eqc.govt.nz/canterbury-earthquakes/land-claims/complex-land-claims/increased-risk-of-flooding.

Fischer-Smith, R. (2013). The earthquake support subsidy for Christchurch's small and medium enterprises: Perspectives from business owners. *Small Enterprise Research* 20(1): 40–54.

Giovinazzi, S., T. Wilson, C. Davis, et al. (2011). Lifelines performance and management following the 22 February 2011 Christchurch earthquake, New Zealand: Highlights of resilience. *Bulletin of the New Zealand Society for Earthquake Engineering* 44(4): 402–417.

GNS Science (2011). June 13, 2011—Large earthquakes strike south-east of Christchurch. *GeoNet*. http://info.geonet.org.nz/display/home/2011/06/13/Jun+13+2011+-+Large+earthquakes+strike+south-east+of+Christchurch.

Hatton, T., E. Seville, and J. Vargo (2012). *Improving the Resilience of SMEs: Policy and Practice in New Zealand* (No. 2012/012). Christchurch, NZ: Resilient Organisations. http://resorgs.org.nz/images/stories/pdfs/apec%20sme%20resilience%20policy%20final%20report.pdf.

ICNZ (2012). *Insurance Council Review 2012*. Wellington, NZ: Insurance Council of New Zealand. http://www.icnz.org.nz/wp-content/uploads/icnz-annual-review-2012.pdf.

IRD (2014). *The Impact of the Canterbury Earthquakes on Small and Medium Enterprises Adverse Events Longitudinal Study*. Wellington, NZ: Inland Revenue Department. http://www.ird.govt.nz/resources/6/8/6888ba4a-437b-4de5-bc49-98e6dc8faf0c/r-and-e-adverse-events-year-2-admin.pdf.

Kachali, H. (2013). *Key Elements of Sectoral Recovery and Resilience after the Canterbury Earthquakes: a System Dynamics Approach* (doctoral dissertation). Christchurch, NZ: University of Canterbury.

King, A., D. Middleton, C. Brown, et al. (2014). Insurance: Its role in recovery from the 2010–2011 Canterbury earthquake sequence. *Earthquake Spectra* 30(1): 475–491.

Liu, M., S. Giovinazzi, R. MacGeorge, and P. Beukman (2013). *Wastewater Network Restoration Following the Canterbury (NZ) Earthquake Sequence: Turning Postearthquake Recovery into Resilience Enhancement*. Reston, VA: International Efforts in Lifeline Earthquake Engineering (Monographs No. 38), doi: 978-0-7844-1323-4.

MBIE (2013). *Housing Pressures in Christchurch: A Summary of the Evidence, 2013*. Wellington, NZ: Ministry of Business, Innovation, and Employment. http://www.dbh.govt.nz/userfiles/file/publications/sector/pdf/christchurch-housing-report.pdf.

Muir-Wood, R. (2012). The Christchurch earthquakes of 2010 and 2011, in C. Courbage and W.R. Stahel (eds) *The Geneva Reports: Risk and Insurance Research No. 5*. Geneva: The Geneva Association, 93–106. www.genevaassociation.org.

NIU (2013). *Infrastructure 2013: National State of Infrastructure Report*. Wellington, NZ: National Infrastructure Unit Implementation of the 2011 Plan. http://www.infrastructure.govt.nz/plan/2011implementation/2013report.

Norman, S. (2006). New Zealand's holistic framework for disaster recovery. *The Australian Journal of Emergency Management* 21(4): 16–20.

NZ Police (2012). *Christchurch Earthquake: List of Deceased*. Wellington, NZ: New Zealand Police. http://www.police.govt.nz/major-events/previous-major-events/christchurch-earthquake/list-deceased.

NZ Treasury (2011). Economic and fiscal impacts of the Canterbury earthquakes. *2013 Budget and Fiscal Update*. Wellington, NZ: The New Zealand Treasury: Kaitohutohu Kaupapa Rawa, 95–101.

NZCID (2013). *Insights into Canterbury: Findings of Research on the Canterbury Earthquakes Recovery*. Auckland, NZ: New Zealand Council for Infrastructure Development. http://www.nzcid.org.nz/Category? Action=View&Category_id=217.

NZLC (2014). *Overview of the New Zealand Lifelines Committee*. Wellington, NZ: Ministry of Civil Defence and Emergency Managment: New Zealand Lifelines Committee. http://civildefence.govt.nz/cdem-sector/lifeline-utilities/new-zealand-lifelines-committee/.

Olshansky, R.B. and L.A. Johnson (2010). *Clear as Mud: Planning for the Rebuilding of New Orleans*. Chicago, IL: American Planning Association.

Orion (2009). *Asset Management Plan*. Christchurch, NZ: Orion New Zealand Ltd. http://www.oriongroup.co.nz/publications-and-disclosures/asset-management-plan.aspx.

Orion (2014). Q&As about Orion's post earthquake pricing, resilience and reliability, in *About Orion*. Christchurch, NZ: Orion New Zealand Ltd. http://www.oriongroup.co.nz/downloads/Key_questions_and_answers.pdf.

Platt, S. (2012). *ReBuilDD Field Trip February 2012: Reconstruction in New Zealand Post 2010–11 Christchurch Earthquakes*. Christchurch, NZ: Cambridge Architectural Research Ltd. http://www.carltd.com/sites/carweb site/files/Reconstruction%20New%20Zealand%20Post%202010-11%20 Christchurch% 20Earthquakes_0.pdf.

Recover Canterbury (2012). *Lessons from Recover Canterbury*. Christchurch, NZ: Recovery Canterbury: Supporting Canterbury Business. http://www.rec overcanterbury.co.nz/.

Rogers, N., K. Williams, M. Jacka, et al. (2014). Geotechnical aspects of disaster recovery planning in residential Christchurch and surrounding districts affected by liquefaction. *Earthquake Spectra* 30(1): 493–512, doi:10.1193/021513EQS029M.

Rotimi, J.O. (2010). *An Examination of Improvements Required to Legislative Provisions for Post-disaster Reconstruction in New Zealand* (doctoral dissertation). Christchurch, NZ: University of Canterbury.

Rotimi, J.O., S. Wilkinson, K. Zuo, and D. Myburgh (2009). Legislation for effective post-disaster reconstruction. *International Journal of Strategic Property Management* 13(2): 143–152.

SCIRT (2014a). *SCIRT Learning Legacy: The Lessons from SCIRT*. Christchurch, NZ: UC Quake Centre. http://www.quakecentre.co.nz/SCIRT-Learning-Leg acy-__I.678.

SCIRT (2014b). *The Value of SCIRT (Interim Report): Abridged Version*. Christchurch, NZ: Stronger Christchurch Infrastructure Rebuild Team. http:// strongerchristchurch.govt.nz/sites/strongerchristchurch.govt.nz/files/2013-12-11TheValue of SCIRTAbridged_0.pdf.

Seville, E., D. Brunsdon, and J. Hare (2014). Your building's resilience. *Build* 141(April/May 2014): 51–53.

Shelton, R. (2014). Resilient building design. *Build* 141 (April/May 2014): 48–50.

Spence, R. (2004). Risk and regulation: Can improved government action reduce the impacts of natural disasters. *Building Research and Information* 35(5): 391–402.

St. Clair, M. and D. McMahon (2011). *Canterbury Fact Finding Project Consideration of liquefaction and lateral spreading hazards in the zoning and development of: Part A: The eastern suburbs of Christchurch from 1977 to 22 Feb 2011 and Part B: Brooklands, Kaiapoi and Kairaki/The Pines*. Christchurch, NZ: Young Cooper Ltd and Resource Management Group Ltd. for the Ministry of the Environment.

Statistics New Zealand (2013). *2013 Census*. Wellington, NZ: Statistics New Zealand Tatauranga Aoteroa. http://www.stats.govt.nz/Census/2013-census. aspx.

Stevenson, J.R. (2014). *Organisational Resilience after the Canterbury Earthquakes: A Contextual Approach* (Unpublished doctoral dissertation). Christchurch, NZ: University of Canterbury.

Stevenson, J.R., E. Seville, H. Kachali, and Z. Whitman (2011). *Post-Disaster Organisational Recovery in a Central Business District Context: The 2010 and 2011 Canterbury Earthquakes* (No. 2011/03). Christchurch, NZ: Resilient Organisations. http://resorgs.org.nz/images/stories/pdfs/Resilient_Organisations_Report_CBD%20Recovery_Final.pdf.

Stirling, M., M. Yetton, J. Pettinga, et al. (1999). *Probabilistic Seismic Hazard Assessment and Earthquake Scenarios for the Canterbury Region, and Historic Earthquakes in Christchurch* (No. U99/18). Lower Hutt, NZ: Institute of Geological and Nuclear Sciences Ltd.

Swiss Re (2012). Swiss Re's sigma on natural catastrophes and man-made disasters in 2011 unveils USD 116 bilion in insured losses and record economic losses of USD 370 billion. *Swiss Re: News Releases*. http://www.swissre. com/media/news_releases/nr_20120328_sigma_disasters_2011.html.

Taylor, J.E. (2013). *The Christchurch Earthquake Sequence: Government Decision-Making and Confidence in the Face of Uncertainty* (Masters Thesis). Kingston, Ont.: Queen's University. https://circle.ubc.ca/ bitstream/handle/2429/45214/ubc_2013_fall_taylor_josh.pdf?sequence=4.

Wilkinson, S. and Y. Chang (2011). Hasty code changes may slow recovery. *Build* 124 (June/July 2011): 14.

Wotherspoon, L.M., R.P. Orense, M. Jacka, et al. (2014). Seismic performance of improved ground sites during the 2010–2011 Canterbury earthquake sequence. *Earthquake Spectra* 30(1): 111–129, doi:10.1193/082213EQS236M.

Zandi, M., S. Cochrane, F. Ksiazkiewicz, and R. Sweet. (2006). Restarting the Economy, in E.L. Birch and S.M Wachter (eds) *Rebuilding Urban Places After Disaster: Lessons from Hurricane Katrina*. Philadelphia: University of Pennsylvania Press, 104–116.

Chapter 7

Adaptive Strategies of Urban Disaster Recovery Planning

Mark Kammerbauer

Introduction

When disaster impacts cities, planners are required to address two central aims: on the one hand, cities need to recover after the disaster; on the other, urban development that took place before the disaster should continue afterwards. Forms of planning exist for both aims: recovery planning and urban planning. Questions arise as to how similar or different they are, how they complement or contradict each other, and whether it is possible to plan for the rebuilding of existing structures and implement necessary urban improvements at the same time. Lack of coordination between the two can lead to uneven, fragmented, ad hoc recovery that neglects the needs of vulnerable populations. As a result, opportunities for beneficial change that could increase the resilience of communities remain unrealized. This leads to the main research question of this contribution: how do the two forms of planning interact when disasters impact vulnerable populations?

To address the issue, this chapter deals with contradictions between recovery planning and urban master planning in the case of Hurricane Katrina (US) and the resulting urban mega-disaster in New Orleans. While recovery planning is intended to enable a "return to normal," urban master land use planning is aimed at an alteration or transformation of a given situation to achieve community goals. In the case of New Orleans after Hurricane Katrina in 2005, the Citywide Recovery Plan attempted to address the immediate aftermath of the disaster, while the new Master Plan focuses on growth and improvement. It is argued that both planning models have been inadequate to meet the needs in areas of the city with concentrations of highly vulnerable residents. As a result, these areas lag behind significantly in recovery efforts. Here, the conjunction of planning and the vulnerability of the city's population seems critical.

In the following discussion, socio-spatial perspectives relative to vulnerability, urban planning, and disaster recovery planning provide a theoretical basis for this chapter. The planning context will be outlined and a methodological discussion will describe the employed case study approach. Within a case study area that experienced massive population loss and low rebuilding rates, New Orleans' Lower Ninth Ward, data were collected via mixed-method quantitative and qualitative research. The results show how vulnerability and lack of capacity to rebuild are related to the city's recovery plan and urban master plan. The conclusion points out the relevance of contradictory planning aims in relation to repopulation and rebuilding in the city. In particular, the density of impacted neighborhoods emerge against the background of existing urban shrinkage processes, which calls into question the master plan goals. While recovery plans and related institutions in New Orleans adapted to these circumstances through the definition of Policy Areas and the Lot Next Door program, the new Master Plan permitted potentially unsustainable development in low-density neighborhoods.

This chapter intends to further integrate the existing body of knowledge on disasters in general, and Katrina in particular, with concepts of urban planning to identify reasons for problems observed in post-disaster recovery that can be traced back to the local planning context and related planning aims. It is therefore oriented toward researchers interested generally in cities and disasters, and specifically in long-term recovery and reconstruction. Resulting fact-based planning recommendations may enable a balance between existing conditions and necessary adaptations and improvements to support rapid, yet sustainable, recovery from urban mega-disasters. The approach is applicable to cases of urban disaster in the USA and adaptable to other (global) cases of coastal and peri-coastal metropolises subject to (hydro-meteorological) risk by taking the particular spatial and social context into account.

Socio-Spatial Approaches to Planning

The two forms of planning are outlined and a comparison illustrates their differences. This approach reinforces the need to consider vulnerability in both forms of planning and stresses that the institutional dimension plays a particular role in relation to the vulnerability of impacted populations. This provides the basis for a discussion of the need for coordination between the two forms of planning to enable just and sustainable recovery strategies: what matters in urban master planning matters in recovery planning as well. It also serves to produce a set of questions relevant to the empirical research.

Socio-spatial perspectives in urban studies and disaster research provide a theoretical basis for research on planning in the context of both cities and disaster, and also their relation to social and demographic aspects. These perspectives state that "spatial or environmental and locational considerations are always part and parcel of everyday social relationships" (Gottdiener and Hutchison, 2006: xv). This is based on a political economy approach toward the production of space, associating economic development "with important factors in the social and political change of metropolitan space" (Gottdiener and Hutchison, 2006: 86). These factors relate to particular social and demographic aspects, including "social class, race, gender, ethnicity, age, family status, and religion" (Gottdiener and Hutchison, 2006: 1) and refer to spatial aspects such as "the clustering of homes according to family income."

The concept of vulnerability can serve to explore the interrelations between the social and institutional dimensions of disaster. Vulnerability is defined in different ways according to different concepts, methods, and systems. Various disciplines have their own meanings of vulnerability. Originating from the perspective that research on disasters oriented on physical hazard is inadequate to explain its social dimension, vulnerability is considered multidimensional, differential, and dynamic (Birkmann, 2006: 10–12). Bohle (2001) proposes a double structure of vulnerability in order to discuss its causes. "External" and "internal" sides of vulnerability are identified, where the external side refers to a structural context and the internal side includes coping strategies that people have access to. Further, the two sides can be viewed as dialectically interactive. Here, vulnerability is considered to be related to control over assets, which may contribute to resilience from environmental risks based on the structural, political, economic, and societal aspects of the region under observation (Bohle, 2001: 3).

Planning can be described as an activity that has the capacity "to predict the consequences of its actions" (Campbell and Fainstein, 2001: 2) and to adapt "to changes in the city and region, which in turn are transformed by planning and politics" (Campbell and Fainstein, 2001: 4). Its aim is "to alter the existing course of events" (Campbell and Fainstein, 2001: 6) by addressing "social, economic, and environmental challenges" (Campbell and Fainstein, 2001: 8). Planning is context-specific, as "problems and conditions of planning are not everywhere the same" (Friedmann, 2003: 6). Contemporary urban planning includes "participatory, broad-based strategies for managing urban change" (Kaiser and Godshalk, 2007: 367). It originated in the notion that master plans "should be focused on long-term physical development" (Kaiser and Godshalk, 2007: 366). Comprehensive plans can include information on "amount, pace, location, and costs of growth as policy choices" (Kaiser and Godshalk, 2007: 374). In this regard, a master plan for future land use can serve as a framework "for community consensus on future growth" (Kaiser and Godshalk, 2007: 366) in relation to "effective implementation under turbulent conditions" (Kaiser and Godshalk, 2007: 367). Plans can also serve as a basis for applying for grants from the federal government for community development and other purposes (Kaiser and Godshalk, 2007: 370).

A brief review of relevant concepts serves to define key elements of planning for disaster recovery. Disasters can be defined as consensus crises as opposed to conflict crises (Tierney, 2006). US legislative definitions of natural disaster include "hurricane, tornado, storm, high water, wind-driven water, tidal wave, tsunami, earthquake, volcanic eruption, landslide, mudslide, snowstorm, or drought" (Stallings, 2005: 240–241). Government authorities are required to respond when a disaster occurs and causes a "breakdown of public order and safety" (Dombrowsky, 1998: 20). The "trigger determines the measure" (Dombrowsky, 1998: 20) of institutional intervention; that is, who responds and how they respond depend on the nature of the disaster. Emergency management includes the skills required to deal with harmful events such as disasters, and it is considered a public service (McEntire, 2006: 169). Drabek points out that coordination is the paramount principle of emergency management. It is defined as "those sets of behaviors through which the complex network of interrelated events are maintained" (Drabek, 2006: 218). This becomes important between the emergency management tiers of government "within local, state, and federal agencies" (Drabek, 2006: 219). However, Drabek notes that the "coordination relevant to disaster within US Communities is very different from that applicable to other settings in other times" (Drabek, 2006: 219).

Within the disaster cycle (Smith and Birkland, 2012: 147–170), emergency management can be structured into phases, including planning and preparedness, mitigation, response, and recovery. In the US, the Robert T. Stafford Disaster Relief and Emergency Assistance Act (Stafford Act) serves as the legislative basis for emergency management. It defines how the Federal Emergency Management Agency (FEMA) and other institutions operate during and after disaster. Governmental assistance is supposed to enable the "reconstruction and rehabilitation of devastated areas" (42 U.S.C. 5121–5207. 1). Further, the Stafford Act calls for communities to return to "normal pattern[s] of life as soon as possible" (42 U.S.C. 5121–5207. 48). For this purpose, recovery funding is provided through supplemental appropriations approved by the US Congress as well as continuing programs such as FEMA's Hazard Mitigation Grant Program and the HUD Community Development Block Grant-Recovery Program (Smith and Sandler, 2012).

Urban disaster is viewed as an interruption of existing urban development processes, including growth. Research on the recovery of cities after disasters finds that "the time needed for reconstruction reflects not only the amount of damage and the available recovery resources, but also pre-disaster trends" (Haas, Kates and Bowden, 1977: xxvi). Recovery is considered complete when the evacuated populations have returned, debris has been removed, and infrastructure and services have been restored. Further, recovery can be structured into periods including restoration, replacement reconstruction, and betterment reconstruction, each requiring successively longer amounts of time in relation to the scale of disaster. The latter may be oriented toward future urban development (Haas et al., 1997: 3). Overly ambitious plans may cause delays to the recovery process; enabling a transition between the two reconstruction periods is recommended. Recovery planning can also lead to inequalities when "official priorities provide disproportionate assistance to certain kinds of people and certain kinds of places" (Vale and Campanella, 2005: 353). Thus, rebuilding may involve "value-laden questions about equity [and] the needs of low-income residents" (Vale and Campanella, 2005: 13).

While the vulnerability of the population of New Orleans is now widely recognized, it is still necessary to study its interrelation with the aims of recovery plans and programs and the master plan in New Orleans after Katrina. The following questions arise: what are the roles of the social and spatial contexts and the institutional and temporal aspects? How does this influence the actual rebuilding and reconstruction? Which opportunities arise? In the following discussion, the planning context will be outlined while the empirical case study data offers answers to the aforementioned questions.

Table 7.1 Comparison of recovery planning and master planning

	Recovery planning	Urban master planning
Aim	Reconstruction and rehabilitation	Change, growth
Spatial context	Devastated areas	Urban areas
Social context	Disaster	City
Institutional aspect	Federal government provides funding to impacted communities for recovery plans and programs	Local community responsible for planning, may apply for federal grants
Time aspect	As soon as possible	Long-term

Planning Context

After Hurricane Katrina impacted the Gulf Coast of the US in August of 2005, congressional emergency supplemental appropriations provided funding for recovery, while the impacted states were responsible for action plans (Czerwinski, 2008: 9–12). Funding was further increased due to Hurricane Rita, which struck in September of 2005. Several key issues compromised recovery for individual homeowners: the immense degree of housing damage, uninsured losses or insufficient funding based on insurance payouts, lack of private resources and loans. Based on the action plan submitted by the state of Louisiana, the state's central rebuilding effort, titled the Road Home program,[1] faced numerous problems, including a difficult and slow application process, budget deficits, and legal and

1 After suffering massive housing damage due to Hurricanes Katrina and Rita, the state of Louisiana created a coordinating institution, the Louisiana Recovery Authority (LRA). Use of funds as intended by the LRA was subject to approval by state legislature. The LRA developed the Road Home program, which was introduced in 2006 with the intention of "encouraging homeowners to (…) begin rebuilding" (Czerwinski, 2007: 7) and "re-establishing their lives in Louisiana" (Road Home n.d.: n.p.). The program's primary focus was on owner occupants who could apply for grants based on property values "before the storms and the amount of damage that was not covered by insurance or other forms of assistance" (Czerwinski, 2007: 7). The Road Home program was continually altered and was comprised of four different options for homeowner recovery assistance: repair, rebuild, relocate, or sell. Preconditions included that applicants were owner-occupants of "a single or double unit structure" (Road Home, n.d., n.p.) as the "main residence at the time of the hurricane" (ibid.) and that FEMA had categorized these structures as either "destroyed" or with "major" or "severe" damage. Applicants could receive a maximum grant of $150,000. Other forms of support, such as insurance monies, were deducted from the total amount. The

regulatory conflicts which caused changes while the program was running. The most significant change was the inclusion of a buyout option, which enabled applicants to sell their properties and homes to the state instead of rebuilding them. This essentially counteracted the program's original aim.

Parallel to these federal and state initiatives, planning in the city slowly arose from difficult beginnings. Initial recovery planning proposals in New Orleans after Katrina included the "Action Plan to Rebuild New Orleans" of the Bring New Orleans Back Commission (BNOB), and the "New Orleans Neighborhoods Rebuilding Plan" presented by the City Council. Eventually, the "Unified New Orleans Plan" (UNOP) was developed as a collaborative effort and outlined the aims of the "Citywide Strategic Recovery and Redevelopment Plan" ("Citywide Plan"). It focuses on reconstruction after disaster and is defined as a type of plan different than a master land use plan. The Citywide Plan was created by a "Citywide Team" that included "local urban planning practitioners and university professors" (City of New Orleans, 2009: 13) who collaborated with district planners from nationally recognized architectural firms to create district and neighborhood plans. In the course of the planning processes and citizen participation taking place after Katrina, the new district and neighborhood plans were integrated within a new master plan with a 20-year timeframe. In 2007, the City Charter was amended, thus providing the basis for the City Planning Commission to create the 2030 Master plan and the Comprehensive Zoning Ordinance in relation to "the city's capital improvement plan [and] a neighborhood participation system for land use decision making" (City of New Orleans, 2009: 13).

The Citywide Plan, on the other hand, describes the priorities of reconstruction in the city in relation to funding within a recovery timeframe spanning five to 10 years. It serves to guide the recovery process "in a rational way that creates stability and paves the way for future growth and prosperity" (City of New Orleans, 2009: 9). The aim is to achieve a "quality of life in New Orleans [that] is back to—or better than—what it was before Katrina" (City of New Orleans, 2009: 10). However, the UNOP also acknowledges that "it cannot be assumed that everyone will want to, or be able to, return to the City" (City of New Orleans, 2009: 9). This also relates to lacking "financial or manpower assets to fix everything at once" (City of New Orleans, 2009: 10) and leads to a prioritization of recovery efforts according to Policy Areas A, B, and C. Policy Area A features low risk and a high degree of repopulation; Policy Area B displays moderate risk and repopulation; Policy Area C is confronted with a very high risk of flooding and very low rates of repopulation.[2]

Road Home program included a component for the owners of rental properties, including "large developers and small property owners." These were implemented only after delays.

2 City of New Orleans, 2007: 65. To address recovery problems of hard-hit areas of the city, the Citywide Plan was supplemented by the Citywide Recovery Implementation Strategy (Recovery Strategy), created by the Office of Recovery Development and Administration (ORDA) and adopted by the City Council in 2007. The Recovery Strategy

In 2009 approximately 45,000 Road Home program closings took place in New Orleans, and 40,000 of those decided to rebuild (GNOCDC, 2009: 13). According to 2010 estimates, roughly 4,500 properties "were sold to the state via the Road Home program" (Plyer, 2010: 3). The properties were acquired by the New Orleans Redevelopment Authority (NORA), which was originally responsible for neighborhood development and clearance of slums. Based on a Master Cooperative Endeavor Agreement, NORA became the city's contractor in the recovery effort in 2008. NORA already had the competence to acquire blighted properties before Katrina. To deal with both blight and the vacancies created by Road Home buyout properties, NORA developed the Lot Next Door Incentive Program (Lot Next Door program). It offers homeowners the opportunity to purchase adjacent properties from NORA in this context (City of New Orleans, 2008: 12). As a result, homeowners were able to both increase the size (and value) of their properties while also contributing to neighborhood repopulation.

By June 2009 the Road Home program had produced 4,600 buyout properties that were unevenly distributed across planning districts. The Lower Ninth Ward or Planning District 8 comprised 776 buyout properties at that time—not the highest quantity, but at the highest rate among city planning districts. After peaking in 1960 at 33,000, population numbers in the Lower Ninth Ward had declined to 19,500 before Katrina. It was particularly hard hit by floods caused by collapsing floodwalls along the Industrial Canal, which intersects the urban fabric. The Lower Ninth Ward is comprised of two identifiable neighborhoods (Holy Cross and Lower 9 neighborhoods, the latter sometimes also called the Lower Nine or Lower Ninth Ward neighborhood) and includes Jackson Barracks, a military facility serving the Louisiana National Guard (situated along the border to St. Bernard Parish). Pre-existing social aspects of the Lower Ninth Ward include a majority African American population, high poverty rates, an above-city average percentage of homeowners, and high levels of disabled and elderly citizens. Dense social networks existed, related to patterns of generational housing and long-term residency.[3]

---

defined Target Recovery Areas in 17 neighbourhoods "in need of urban transformation as the result of pre-Katrina deterioration as well as damage incurred both during the 2005 Hurricane Season" (City of New Orleans and New Orleans Redevelopment Authority, 2008: 1–2).

3 See GNOCDC, 2009; H3 Studio Project Team (2007). "A Framework for Sustainable Resilience in the Lower Ninth Ward. Introduction to the District"; The ACORN Housing/University Partnership (2007). "A People's Plan for Overcoming the Hurricane Katrina Blues."

Case Study Methodology

A case study methodology was selected to collect data in the Lower Ninth Ward in 2007 and 2009. The selected research unit displays extensive housing damage and below average repopulation rates and thus meets the criteria for Planning Area C set by the Citywide Plan. Case studies may draw from "multiple sources of evidence" (Yin, 2009: 18) and are useful when cases are unique or offer "a phenomenon previously inaccessible to social science inquiry" (Yin, 2009: 48). Communities are possible research units for case studies (Yin, 2009: 12) and can be related to the concept of neighborhood in urban space, as "neighborhood is often the place where people find community" (Williams, 1985: 34). Researching communities can "provide evidence of specific links to socio-spatial organizations in the area" (Gottdiener and Hutchison, 2006: 195–196).

To research these links between the vulnerability of the community and the plans and programs intended for its recovery, the methodology combines qualitative and quantitative methods within an integrated research approach. Points of origin are qualitative ethnographic interviews, followed by a quantitative questionnaire-based survey. The results were further investigated by qualitative semi-structured expert interviews. The focus on the case study object enables the triangulation of these mixed methods, and the individual methods complement each other (Flick, 2012: 43–44). The aim is to achieve broader and more comprehensive insight into the matter at hand.

The ethnographic interviews were held on site by the author in 2007 in New Orleans's Lower Ninth Ward. The intention was to gain more knowledge on the recovery situation. Three ethnographic interviews with members of nonprofit organizations (Holy Cross Neighborhood Organization, and ACORN, both in New Orleans, and United Way in Houston, Texas) produced information framing the design of the quantitative questionnaire.

The situation on site needed to be considered, since the majority of former residents had not yet returned. As a result, only a partial survey of the overall population was possible. The sampling was non-random and based on a snowball approach. The three members of the non-profit organizations mentioned above established contact with three sample groups (Bryman, 2012: 424), and three survey rounds were conducted as follows: after a meeting of the Holy Cross Neighborhood Organization, after service at St. Maurice Church (via ACORN), both in the Lower Ninth Ward, and at Lakeside Estate housing in Houston (via United Way). Questionnaires were completed while the author was present. The survey produced data on whether respondents were capable or incapable of returning to the Lower Ninth Ward. While the sample is not representative (Schnell, Hill, and Esser, 2011: 298), it points out aspects of the dual structure of vulnerability and its internal and external sides that the qualitative expert interviews were intended to further research.

The semi-structured expert interviews were held in 2009 in New Orleans and Washington, DC with individuals who were identified as experts in a particular field of action, as representative of a particular group, or as members of organizations with specific responsibilities (Flick, 2012: 214–215). The qualitative expert interviews were used to complement the previous quantitative survey. The interview partners were selected in a targeted manner. Questions were open and oriented to the theoretical discussion and recovery and reconstruction efforts at the federal and local levels of government. Interviews were recorded digitally. The interview text is based on verbatim transcripts of the recordings and was used for qualitative content analysis. The analytical categories were developed from the theoretical approach and not derived from interview data. The results were interpreted in relation to the research questions for explanatory purposes (Flick, 2012: 409–410).

Results

Quantitative Survey

The quantitative questionnaire was conducted to research vulnerabilities of a population sample group that lived in the Lower Ninth Ward before Katrina. The major share of the sample group had returned to New Orleans after Katrina, while some had not yet been able to return at that time and lived in Houston, a major destination for Katrina evacuees. The questionnaire was designed to enable comparison to the US Census (GNOCDC, 2007) and a larger survey conducted among New Orleans evacuees in Houston (Wilson and Stein, 2007). Seventy-five completed questionnaires were collected and calculated as 100 percent. The questions focused on evacuation and return, housing conditions, and support in recovery. Some similarities to Wilson and Stein's observations are noteworthy, such as the observation that the evacuees who remained in Houston were unlikely to return, mostly "because very few of these respondents owned their homes in New Orleans" (Wilson and Stein, 2007: 7).

The survey also features a majority of individuals who noted their ethnicity as African American. Incomes were below $25,000 (seven respondents noted a household income of less than $15,000 after Katrina), and the majority were renters. Differences with census data and the larger Houston survey include the fact that the author's survey indicated a higher average age of participants (42.5 years) and a higher degree of female respondents. In sum, pre-existing demographic and socio-economic aspects of income and ethnicity can be interpreted as vulnerabilities in the context of disaster:

- most respondents are homeowners who had been able to return;
- most housing was either severely damaged or completely destroyed;
- difficulties emerge in applying for funding for recovery;
- the proportion of renter households increased;

- long-term evacuees are mostly renters with low income; and
- returning individuals require alternative housing while rebuilding damaged homes.

Qualitative Expert Interviews

A staff member of the United States House of Representatives Committee on Homeland Security (Interview 1, Table 7.2), responsible for emergency management issues, discussed the impact of federal legislation on post-Katrina recovery with the author. In the case of Hurricane Katrina, using the CDBG program to provide funding for recovery was considered appropriate, because no new program had to be developed to authorize funding [Cit. 1]. The CDBG program poses problems, yet has benefits: recipients can decide on how to use the grant funds [Cit. 2]. This is also seen as an opportunity for bottom-up decision making [Cit. 3, 4].

Table 7.2 Citations showing federal justification for institutional aspects

Citation	Interview 1
Cit. 1	"the CDBG program is a useful preexisting tool that is able to be funded very quickly"
Cit. 2	"states can use CDBG funding for whatever they want"
Cit. 3	"the CDBG program offers the opportunity to propose plans to the federal government from the ground up"
Cit. 4	"nobody understands the issues on the ground like the locals do"

The Director of Real Estate Strategy for the New Orleans Redevelopment Authority (Interview 2) offered a detailed view of the interactions between recovery funding, programs, and plans, as well as their impact on those who apply for funding. The fact that the state of Louisiana received the CDBG funding for recovery is not considered beneficial for rebuilding the city. It is considered highly important that the funding methods match the local disaster context [Cit. 5]. Flexible interpretation is necessary for the recovery of the city, which the state may not consider necessary [Cit. 6], because of the added burden of having to repay funds [Cit. 7]. In return, this leads to problems with the Road Home program, which affect the recovery of the city [Cit. 8].

Table 7.3 Citations showing local critique of institutional aspects

Citation	Interview 2
Cit. 5	"the rules of a traditional program like CDBG do not fit disaster recovery"
Cit. 6	"you need somebody willing to take legitimate and reasonable interpretation"
Cit. 7	"the state has no incentive ultimately to take an aggressive interpretation (…) because that ensures they will never have to repay a dime of this money"
Cit. 8	"the state management of funds leads to the Road Home program becoming a fundamentally unfair program to New Orleans"

What are the particular reasons for the problems with the city's recovery in regard to the Road Home program? For one, it does not adequately address the ratio of homeowners to renters in the city [Cit. 9]. At 46.5 percent, the share of homeowners in New Orleans before Hurricane Katrina was lower than the statewide figure of 67.9 percent (Louisiana Recovery Authority, 2007: 4). Further, the Buyout option is considered adverse to the intention of rebuilding the city [Cit. 10, 11]. This is based on the discrepancy between the funding provided and actual rebuilding costs on the one hand [Cit. 12, 13], and on the reluctance to return after potentially having resettled elsewhere while waiting for the approval of recovery funding applications [Cit. 14].

Table 7.4 Citations showing impact of Road Home program

Citation	Interview 2
Cit. 9	"the Road Home has a disproportionate share of the rental-type programs"
Cit. 10	"the Buyout option results in no effective procedures to guarantee rebuilding"
Cit. 11	"now you could take the Road Home money and walk away"
Cit. 12	"if you lost a 50,000 dollar house, it will still cost you 150,000 dollars to rebuild"
Cit. 13	"if you only give someone 50,000 dollars for their house they lost that was in a poor neighborhood, they might have sold the property to the state"
Cit. 14	"if it took you three years to get money to rebuild your house, and you have just spent three years living somewhere else, … re-uprooting your life to come back is not necessarily what is going to happen"

The Citywide Plan adapted out of necessity to this situation by formulating different recovery policies and Policy Areas. This is also due to the fact that a recovery plan didn't exist when Katrina struck, preventing timely coordination with the Road Home program [Cit. 15]. Policy Areas are differentiated according to degree of destruction and repopulation. The most successful neighborhoods belong to Policy Area A, comprising the historic parts of the city that received little damage. The most difficult neighborhoods belong to Policy Area C, where low return rates limit the effectiveness of NORA's Lot Next Door program [Cit. 16, 17]. Policy Area C denotes parts of the city experiencing below-average repopulation and rebuilding, thus calling for targeted recovery initiatives. This is related to pre-existing conditions in these areas [Cit. 18]. As a result of low repopulation, limited or targeted recovery efforts are defined in relation to available resources [Cit. 19, 20].

Table 7.5 Citations showing impact of the Citywide plan

Citation	Interview 2
Cit. 15	"what you would have wanted would have been for this Road Home Program to match your recovery plan"
Cit. 16	"until it gets from that sort of incredibly vulnerable 50–80 percent unoccupied level, a neighborhood is more likely of going down that it is of going up"
Cit. 17	"there are urban areas in the city that are in tremendous states of decay"
Cit. 18	"they display historic poverty combined with massive storm damage, population loss, economic loss"
Cit. 19	"in these areas, recovery is relatively narrowly targeted: you pick a few blocks, a few zones"
Cit. 20	"otherwise you will quickly dilute whatever is feasible, whatever energy those areas have"

The new Master plan and Comprehensive Zoning Ordinance are viewed critically due to their emphasis on growth, which appears to contradict historic developments and the actual recovery situation. New Orleans was a shrinking city before Katrina and struggled with slow repopulation in particular areas during the recovery phase [Cit. 21]. The Master Plan also does not address the difficult recovery in Policy Area C, where density is decreasing [Cit. 22]. Even more, it promotes large residential developments within impacted lower density neighbourhoods surrounding the city centre [Cit. 23]. Adapting development policies to match the existing contexts is considered a more sensible approach [Cit. 24]. (See also Olshansky and Johnson, 2010.)

Table 7.6 Citations showing impact of Master plan

Citation	Interview 2
Cit. 21	"basically the Master plan is projecting very high levels of growth for the city, which runs contrary to the last 45 years"
Cit. 22	"this ignores the facts that in a number of our neighborhoods we're facing a difficult challenge right now, because they're 50 percent occupied"
Cit. 23	"the Master plan allows multifamily development outside the downtown"
Cit. 24	"for reasons of de-densification for integration and everything else, a single family plotting of people is a much more resilient pattern"

Figure 7.1 Lower Ninth Ward, typical shotgun house
Source: Author's photograph.

The lack of coordination between state and local programs/plans led to a situation where the city's recovery plan needed to adapt to the actual situation and depart from the 'return to normal' paradigm.[4] However, shrinkage was not adequately acknowledged by the Master Plan. This caused problems, particularly in hard-hit neighborhoods belonging to Policy Area C with existing social disparities such as low incomes and low property values. These are reflected in the vulnerability of residents and the lack of capacity of low-income renters to return, as well as the need for interim housing among those who did return. In sum, this magnified the problem of low return and low rebuilding rates.

Figure 7.2 Lower Ninth Ward, multifamily development
Source: Author's photograph.

4 Differences between planning for recovery and for 'normal' urban development are pointed out. For instance, independent planning commissions exist so "that politicians aren't able to dictate … waivers and variances" (Interview 2). The precondition for this is "insulat[ing] as many of these boards and commissions from the political process." During disaster recovery, however, "that is all upside-down."

Figure 7.3 Lower Ninth Ward, small rental property
Source: Author's photograph.

Conclusion

The following contradictions between recovery planning and master planning are indicated by the empirical data, impairing the recovery of the Lower Ninth Ward in New Orleans after Hurricane Katrina:

- aims: the Citywide Plan addresses shrinkage while the Master Plan advocates growth.
- spatial context: the pre-existing and post-disaster conditions in the city are interpreted differently, and the Master Plan concentrates on density where controlled de-densification appears more sensible.
- social context: neither form of planning adequately addresses the vulnerability of the impacted population; the Road Home program for individual homeowner recovery exacerbated this situation due to its focus on homeowners.
- institutional aspect: the federal government assumed that state management of funds would enable an implementation of program goals matching local needs, yet in the case of the city this led to a lack of coordination: recovery planning and institutions in the city were required to adapt to the situation

created by the Road Home program, thus leading to the targeted (that is limited) strategies of the Citywide Plan.
- time aspect: due to delays in providing necessary funds to applicants (caused by post-disaster planning issues), quick recovery did not occur: residents requiring interim housing emerged as a problem and Road Home applicants tended toward choosing the buyout.

While the Citywide Plan demonstrates a departure from the return to normal paradigm of recovery planning, it doesn't seem that beneficial change was effectively addressed. In fact, while a buyout may be sensible for residents of neighborhoods subject to continued risk, it led to ad hoc and fragmented de-densification. An opportunity for controlled and sustainable shrinkage remained unused. Neither was the vulnerability of impacted residents addressed adequately. The Lower Ninth Ward exemplifies this situation pre-storm – a majority African-American population, below-average incomes, and homeowners with low property values. The situation is most pronounced in the Lower 9 neighborhood, partially below sea level and subject to increased risk, as compared to the historic Holy Cross neighborhood, situated along the natural levee of the Mississippi.

The NORA Lot Next Door program comprises a form of adaptation in recovery planning aimed at fostering infill development. In Policy Area C this becomes extremely difficult due to low repopulation rates and an over-abundance of vacant properties. This situation is further complicated by the aspirational aims for urban growth formulated by the new Master Plan. The large, multi-family development it supports throughout the city may even lead to a concentration of risk. During the still ongoing recovery process, its aims become doubly contradictory: they defy historic shrinkage before Katrina and low repopulation rates in hard-hit areas of the city after the disaster. In sum, these observations indicate the weak link between institutional aspects of recovery planning and urban master planning on the one hand, and vulnerability on the other.

For recovery to be effective, institutional support and funding must be provided quickly.[5] Recovery planning and master planning, while comprising two different forms of (urban) planning, require coordination of planning goals in relation to existing conditions. Therefore, the author proposes the following planning recommendations: planning before disaster is beneficial, as it enables coordination between recovery programs and plans (see also Smith and Sandler, 2012). State-led recovery funding and programs must take into account that cities are comprised of both homeowners and rental populations: strategies need to be adjusted accordingly. To assist both homeowners and renters in low-density neighborhoods, targeted and early support for small rental properties should be provided, particularly where infill development appears more sustainable than large multi-family development. Rebuilding what existed before the storm

5 A balance between speed and deliberation is advocated by Olshansky (2005).

(reconstruction) and improving this situation (betterment) need to be balanced carefully to avoid contradictions between shrinkage and growth.

While the empirical data refer to an American example, the theoretical basis, the methodological approach, and the sociological and community-oriented methods are applicable in other contexts and similar (potential) cases. Due to global urbanization processes and the potential dangers of (anthropogenic) climate change and sea level rise, plenty of coastal and peri-coastal metropolises come to mind in this regard. The pre-existing spatial and social contexts of the city constitute a significant influence on recovery planning and urban master planning alike, its aims, and its institutional aspects in relation to the time required for implementation. In the case of urban disasters, to value what exists in the city becomes a matter of survival during recovery after disaster.

References

42 U.S.C. 5121–5207 (2011). *Robert T. Stafford Disaster Relief and Emergency Assistance Act, as amended, and Related Authorities*. http://www.fema.gov/pdf/about /stafford_act.pdf.

Birkmann, J. (2006). Measuring vulnerability to promote disaster-resilient societies: Conceptual frameworks and definitions, in J. Birkmann (ed.) *Measuring Vulnerability to Natural Hazards: Towards Disaster Resilient Societies*. New York: United Nations University, 9–54.

Bohle, H.-G. (2001). Vulnerability and criticality: Perspectives from social geography. *IHDP Update, Newsletter of the International Human Dimensions Programme on Global Environmental Change* 2(1): 3–5.

Bryman, A. (2012). *Social Research Methods*. New York: Oxford University Press.

Campbell, S. and S. Fainstein (2003). Introduction: The structure and Debates of Planning Theory, in S. Campbell and S. Fainstein (eds) *Readings in Planning Theory*. Oxford: Blackwell, 1–16.

City of New Orleans (2007). The Unified New Orleans Plan (UNOP). Citywide Strategic Recovery and Rebuilding Plan. http://planning.uno.edu/Lakefront/docs/UNOP-FINAL-PLAN-April-2007-15744.pdf.

City of New Orleans (New Orleans City Planning Commission, New Orleans Office of Recovery and Development Administration, New Orleans Master Plan and Comprehensive Zoning Ordinance Consultant Team) (2009). *New Orleans Master Plan and Comprehensive Zoning Ordinance. Draft 21 March 2009*. New Orleans, LA: City of New Orleans. http://www.nolamasterplan.com.

City of New Orleans (New Orleans Redevelopment Authority) (2009). *Master Cooperative Endeavor Agreement between the City of New Orleans and the New Orleans Redevelopment Authority*. New Orleans, LA: City of New Orleans. http://www.noraworks.org/PDFs/NORA_CNOCooperativeEndeavor Agreement.pdf.

Czerwinski, S. (2008). "Gulf Coast rebuilding: Preliminary observations on progress to date and challenges for the future." Testimony before the Subcommittee on Disaster Recovery, Committee on Homeland Security and Governmental Affairs Committee, US Senate. http://www.gao.gov/new.items/d07574t.pdf.

Dombrowsky, W. (1998). Again and again: Is a disaster what we call a disaster, in E. Quarantelli (ed.) *What is a Disaster: Perspectives on the Question*. New York: Routledge, 19–30.

Drabek, T. (2006). Community processes: Coordination, in H. Rodriguez, E. Quarantelli, and R. Dynes (eds) *Handbook of Disaster Research*. New York: Springer Science+Business Media, 217–233.

Flick, U. (2012). *Qualitative Sozialforschung. Eine Einführung (5. Auflage)*. Hamburg: Rowohlt.

Friedmann, J. (2003). Toward a non-Euclidian mode of planning, in S. Campbell and S. Fainstein (eds) *Readings in Planning Theory*. Oxford: Blackwell, 75–82.

GNOCDC (2007). Maps. http://www.datacenterresearch.org/.

GNOCDC (2009). The New Orleans index. http://GNOCDC.org/NewOrleansIndex/index.html.

Gottdiener, M. and R. Hutchison (2006). *The New Urban Sociology*. Boulder, CO: Westview.

H3 Studio Project Team (2007). A Framework for Sustainable Resilience in the Lower Ninth Ward. Introduction to the District. New Orleans, LA: H3 Studio Project Team. http://willdoo-storage.com/Plans/D8/District_08_Lower9th_Chapter_01_Introduction_to_the_District.pdf.

Haas, E., R. Kates, and M. Bowden (1977). *Reconstruction Following Disaster*. Cambridge: MIT Press.

Kaiser, E. and D. Godshalk (2007). Twentieth century land use planning: A stalwart family tree, in R. LeGates and F. Stout (eds) *The City Reader* (4th edition). New York: Routledge, 399-420.

Louisiana Recovery Authority (2007). *Proposed Action Plan for the Use of Disaster Recovery Funds Allocated by P.L. 109–234*. Baton Rouge, LA: LRA. http://www.doa.louisiana.gov/cdbg/dr/plans/ActionPlan2-Approved-07-05-24.pdf.

McEntire, D. (2006). Local emergency management organizations, in H. Rodriguez, E. Quarantelli, and R. Dynes (eds) *Handbook of Disaster Research*. New York: Springer Science+Business Media, 168–182.

NORA (2009). *Community Improvement Planning Areas, New Orleans: New Orleans Redevelopment Authority*. New Orleans, LA: NORA. http://www.noraworks.org/district_8.htm.

Olshansky, R. (2005). How do communities recover from disaster? A review of current knowledge and an agenda for future research, in *46th Annual Conference of the Association of Collegiate Schools of Planning*, pp. 1–19.

Plyer, A. (2010). *Benchmarks for Blight. How does New Orleans Compare to Other Cities in Terms of Unoccupied Residential Addresses?* http://www.GNOCDC.org/BenchmarksForBlight/ index.html.

Schnell, R., P. Hill, and E. Esser (2011). *Methoden der empirischen Sozialforschung.* Munich: Oldenbourg Verlag.

Smith, G. and D. Sandler (2012). *State Disaster Recovery Planning Guide.* Chapel Hill, NC: University of North Carolina at Chapel Hill Center for the Study of Natural Hazards and Disasters.

Smith, G. and T. Birkland (2012). Building a theory of recovery: Institutional dimensions. *International Journal of Mass Emergencies and Disasters* 30(2): 147–170.

Stallings, R. (2005). Disaster, crisis, collective stress and mass deprivation, in E. Quarantelli and R. Perry (eds) *What is a Disaster: New Answers to Old Questions.* Newark, DE: University of Delaware International Research Committee on Disasters, 237–274.

The ACORN Housing/University Partnership (2007). *A People's Plan for Overcoming the Hurricane Katrina Blues.* New Orleans, LA: The ACORN Housing/University Partnership. http://www.aap.cornell.edu/aap/crp/outreach/nopi/upload/Peoples_Plan_for_9th_Ward.Pdf.

The Road Home (2010). *Overview: Building a Safer, Stronger, Smarter Louisiana.* Baton Rouge, LA: The Road Home. http://www.lpb.org/programs/LApublicsquare//LRARoadHome.pdf.

Tierney, K. (2006). Social inequality, hazards, and disasters, in R. Daniels, D. Kettl, and H. Kunreuther (eds) *On Risk and Disaster Lessons From Hurricane Katrina.* Philadelphia, PA: University of Pennsylvania Press, 109–128.

Vale, L. and T. Campanella (2005). *The Resilient City: How Modern Cities Recover from Disaster.* Oxford: Oxford University Press.

Wiliams, M. (1985). *Neighborhood Organizations: Seeds of a New Urban Life.* Santa Barbara, CA: Greenwood Press.

Wilson, R. and R. Stein (2007). *Katrina Evacuees in Houston: One-year Out* (white paper). http://brl.rice.edu/Katrina/White_Papers/White_Paper_9_8_06.pdf.

Yin, R. (2009). *Case Study Research. Design and Methods Applied Social Research Methods Series Volume 5.* Sage: Thousand Oaks.

Chapter 8
Urban Vulnerability and Resiliency to Natural Disasters: An Integrative Tourism Planning Perspective

Hyun Kim and David W. Marcouiller

Introduction

In an interesting insight, Buckle (2006, 88) noted that:

> Resilience and vulnerability are central concepts in understanding the nature of disasters, their impacts, and consequences and methods of dealing with the potential for loss. However, neither concept has been exhaustively explored (despite a clear need for intellectual rigor and conceptual clarity), although much more attention has been given to vulnerability than to resilience.

As a central element of disaster research, this passage highlights the difficulty in assessing resilience and vulnerability given the complex and dynamic nature of human systems, communities, and societies. It is estimated that in the last 10 years, major natural disasters across the globe affected more than three billion people, killed over 750,000 people, and cost about US $600 billion in damages (Birkmann, 2006). These severe disasters have also influenced tourism and related business activity in affected areas (Pezzullo, 2009). Most notably tourism-dependent economies along coastal areas of the United States have confronted serious threats from hurricanes, coastal storms, and environmental catastrophes (Stone et al., 1997; Woosnam and Kim, 2014).

Given this level of human loss and property damage, it is safe to say that our communities could be made more resilient to natural disasters. Natural hazards can have overwhelming short- and long-term impacts on the natural and built environments, ultimately affecting local communities and their social and economic conditions. In this sense, it is worth addressing the concepts associated with a planning-relevant model of disaster outcomes that incorporate community vulnerability and resiliency.

Integrative tourism planning focuses on the ability to pro-actively interact with both supply and demand aspects associated with leisure-based migration (Hall, 2008; Marcouiller, 1997). This unique form of human migration reflects travel to and from destinations with particular leisure interests within two specific yet distinct tourism planning contexts. First and most obvious, stimulating short-term destination travel is the most prevalent element in local and state level public policies that addresses tourism and, from a tourism planning perspective, is reflective of regional boosterism. This fundamentally non-integrative approach to tourism planning has been criticized as being myopically focused on marketing to stimulate increased levels of visitation (Ivars Baidal, 2004). Disasters can dramatically alter these short-term travel patterns while resiliency planning can affect the ability of short-term leisure travel to rebound following an incident using targeted marketing. A recent example of this is the relatively quick rebound in Gulf Coast tourism following the 2010 Deepwater Horizon Oil Spill, driven largely by promotional and marketing subsidies provided by British Petroleum.

The second, and perhaps more important, element of planning for touristic migration is reflected in long-term destination travel to second homes and retirement travel (for examples, see Hall and Müller, 2004; Roca, 2013; Schewe et al., 2012). It is well known that this second contextual element of tourism planning is concentrated in regions endowed with high levels of natural amenities (Chi and Marcouiller, 2012), many of which are prone to natural disasters (Cutter and Emrich, 2006). These regions include oceanic and freshwater coasts (rivers, lakes, and reservoirs), mountains, and other regions with natural features that reflect hedonic motivators for travel. Certainly, resiliency planning as a component of integrative tourism planning can assist in addressing the needs of longer-term leisure migrants and provide one important linkage to broader scale disaster planning.

This chapter is written to provide an overview of the literature with respect to disaster planning, with a specific interest in the effects of disaster events on amenity-based development and urban tourism. Together with the basic premise that disaster effects are fundamental social process responses that require pro-active planning, we develop a conceptual model of disaster loss factors that involves regional exposure and shock within the context of inherent social system vulnerability and resilience. By reviewing the extant knowledge related to natural disasters within a community planning context, we highlight the importance of pro-active and integrative tourism planning and public policy to address disaster vulnerability and resiliency.

Following this introduction, our chapter is organized into four subsequent sections. First, we define disaster vulnerability, resiliency, community loss, and planning responses with an emphasis on social frameworks. This is followed with an overview of recent literature on vulnerability and resiliency within a community planning context. This includes both an acknowledgment and a discussion of existing knowledge that significantly contributes to the formation of disaster response in the context of integrative tourism planning. The next section connects indicators of vulnerability, resiliency, and risk in the development of our integrated conceptual planning model of disaster outcomes associated with vulnerability and resiliency particularly relevant to urban tourism. The final section provides conclusions and relevant policy implications.

vulnerability characteristics, Wisner et al. (2004: 11) pointed out that ility can be measured in terms of damage to future livelihoods and social (class, caste, ethnicity, gender, and immigration status); not just what to life and property at the time of the hazard event. As such, vulnerable ties (groups) are assumed to be those who find it hardest to re-construct lihoods following a disaster. This process makes already vulnerable ties more vulnerable to the effects of subsequent hazard events.

vise, vulnerability is the likelihood of being negatively affected by ion and stress, whether in an environmental or socioeconomic system on and Kasperson, 2005; Phillips and Morrow, 2007). As expected, nerable communities (food-insecure, water-scarce, resource-dependent, and states) are highly exposed and poorly equipped to cope with major ought on by natural disasters. In this regard, the vulnerability of human ts is tied to differing social processes (Cardona, 2004).

context of social frameworks with respect to the relationship between isasters and tourism is a growing focus within research on travel Tourism planning with respect to disaster response provides the context rous studies (Burby and Wagner, 1996; Kelman et al., 2008; Tsai and 10) to examine the negative influence of natural disasters on tourism ns; their focus involves tourists themselves, community residents, sm dependent businesses. Woosnam and Kim (2014) analyzed disaster by looking at visitation patterns of 10 coastal park regions in the ern United States in hurricane and tourism seasons (June to November) year from 1979 to 2004. Results of this research suggest that efforts vulnerability of tourism-based regional economies also necessitate an educe unexpected damage through proactive disaster management plans. on, based on the impact of the 2004 tsunami around coastal regions of n Ocean, Kelman et al. (2008) highlighted the importance of tourism ns affected by natural disasters and emphasized the role of tourists in and disaster risk reduction education. Campiranon and Scott (2007) ie (2009) identify additional factors influencing crisis management in estinations, including cultural differences (power distance, uncertainty e, long or short-term orientation), decision making in crisis management d resources (tourism organization, planning and infrastructure, human).

d, natural disasters can have dramatic negative effects on short- and travel patterns and resources (Woosnam and Kim, 2014). Integrative lanning that accounts for regional disaster vulnerability can influence y of short-term travel to rebound following an incident. Further, long-enity-driven migrations to tourism destinations (second homes and t travel) are essential elements of residential populations that require focus in disaster planning. These populations are often absent, and dential dwellings may be vacant during essential public participation s over off-peak tourism seasons. In this sense, an essential linkage

Understanding "Natural" Disasters within "Soc

How we create our communities and where we c
resilient we are to the damages caused by hazards (
to reduce the impact of natural disasters are fundame
sustainable human settlement. The "crisis" elemen
the sociological sense, can be thought of as the failu
communities in adapting to an environmental ever
not simply the result of an isolated high impact r
be viewed as failures to develop and distribute h
community infrastructure capable of withstandin
such an event. From this perspective, community r
can be regarded as a process by which the groups a
the community attempt to re-establish social netw
elements necessary to return stability to the routine

Based on this sociological sense, Rodriguez and
that "disasters are not caused by the 'natural' envir
social, political, and economic environment[s] and
to prepare for and manage the outcomes of sucl
are better able to prepare for, respond to, and rec
others have a limited capacity to resist and recove
of a hazard event (Vollmer, 2013). Without pro-ac
wisdom is that natural disasters lead to extensive h
losses in the economic wellbeing of the affected p

Community-Level Vulnerability and Resiliency
Natural Disasters and Tourism

Conceptualizing both the event and the social cons
can "have a devastating impact on a communit
functions and altering social relationships" (Ho
disasters become a catalyst for much larger socia
and displace communities, evidenced by the situati
of Hurricane Katrina (Swan and Bates, 2007). Suc
constructed as being outside of ordinary exper
coping mechanisms, disrupting social relations, a
individuals and communities (Miller, 2012).

Disaster Vulnerability, Community Losses, Plannin

The concept of vulnerability to natural disasters r
to deal with the impact of natural hazards, to w
consequences, and to cope with the resulting da

terms o
vulnera
structur
happens
commu
their liv
commu

Like
perturba
(Kasper
many v
small is
shocks
settleme

The
natural
behavio
for num
Chen, 2
destinat
and tou
impacts
southea
for each
to lesse
effort to
In addit
the Indi
in locati
disaster
and Rit
tourism
avoidan
stages,
resource

Inde
long-ter
tourism
the abil
term ar
retireme
specific
their re
process

between proactive disaster planning and integrative tourism planning can be helpful in reducing both the risks to and the vulnerability of short- and long-term visitors in disaster prone tourism destination regions.

Vulnerability is a somewhat subjective term indicating the hypothetical, relative likelihood of suffering negative consequences if exposed to an event (or hazard) of a specific strength (amount of energy) or duration.[1] Vulnerabilities can also be thought of as characteristics that determine the ability to absorb the shock of an event. Presumably, communities can lack sufficient resources to cope with disruptions to livelihoods or ill health that may result from exposure to the hazard. Pelling (2012: 148) points out that the urban poor "suffer from a four-fold burden of environmental risk that includes: 1) local public health hazards; 2) localized industrial pollution; 3) inability to adapt to global environmental change including the local impacts of climate change; and 4) particular vulnerability to natural hazards such as earthquakes and storms due to population density." Community development studies often rely on a livelihood framework to understand "community capitals or assets" (Flora, Flora, and Fey, 2003; Green and Haines, 2012). Encompassing a variety of thematic disciplines (economic, social, human, political, cultural, built and environmental), these community capitals are central to understanding disaster response. The loss or devaluation of settlement, livelihood, assets, social security, and status at the local level from the effects of natural hazards can create new vulnerabilities or risks associated with social discrimination, social exclusion, and violence.

Emphasizing equity in natural disasters, Patterson (2013: 101) proposed that a thorough analysis of the myriad elements that lead to pre-disaster vulnerabilities at the local level are necessary to "design policies, programs, processes, and infrastructure that encompass the needs of all." Measures to assess circumstances in disasters that deepen vulnerabilities, or create new risks and hazards for some more than others, need to be included. Most community policies before and after disasters have created inequitable school reforms, as well as marginalizing redistricting processes and redevelopment (Patterson, 2013: 101).

Regarding social justice in vulnerability considerations, most deliberations about local social and economic equity (equal access to quality education, affordable housing, health care, job opportunities, legitimacy and trust) are incomplete and can result in key natural disaster losses that need to be addressed when developing social justice principles (Brown-Jeffy and Kroll-Smith, 2009; Jha et al., 2013). For instance, racial segregation in New Orleans made Black communities physically vulnerable to natural disasters due to local government neglect (Angel et al., 2012; Miller and Rivera, 2007). It should be noted that social justice connections to disaster vulnerability are a long-standing phenomenon. During the Mississippi River Flood of 1927, which covered a 27,000-square-mile area from Illinois to the Gulf of Mexico and dislocated almost a million people, Smith (2011) pointed to

1 We also must note that regional vulnerability is considered spatially mutable, namely that locality specific elements (coastal or inland) dictate level of vulnerability.

a focused, limited survivability of African American communities due to flawed drainage systems and a lack of political interest in aiding these communities on the part of the White population.

As part of the social and economic conditions at the local level, planning and public policy components are associated with disaster preparedness and mitigation in the disaster management cycle (Schwab et al., 2007). Important contributions (for example, Nelson and French, 2002) point out that careful attention to natural hazards in the preparation of local comprehensive plans (including hazard mitigation plans) can result in a reduction of disaster-related losses. Identifying dimensions of policy recommendation and citizen involvement, Burby (2005: 68) suggested key planning elements[2] in demonstrating that "local government comprehensive plans and the process of preparing them can result in lower exposure to losses from natural disasters." These provide the basis upon which resiliency planning and integrative tourism planning for disasters can unfold.

Having community planning and public policy in place before an event to reduce disaster vulnerability involves several elements, which are outlined in Table 8.1. Typically, these include general planning components (local emergency management plans, local land use planning), development regulations (zoning ordinance), building standards (building code), property acquisition (building relocation), critical public facilities, and taxation information dissemination (e.g., Beatley, 2009; Jenkins and Haynes, 2012; Lyles et al., 2014; McAdoo, 2012; McCormick, 2012; Smucker, 2012; Smucker and Wisner, 2007; Underwood and Held, 2012), and are in accordance with disaster types. As for the non-structural flood mitigation approaches, local land use planning techniques allow communities to be more resilient to flooding as suggested by previous studies (e.g., Etkin et al., 2012; Gruntfest, 2000; Lyles et al., 2014; Schmuck, 2012). In addition, the infrastructure element (structural measures such as dikes, dams, levees, river bank stabilization, and building elevation) in comprehensive plans provides resilient and adequate infrastructure linked with a hazard mitigation strategy to protect infrastructure in storm events, and to ensure quick recovery and use during emergency situations (e.g., Beatley, 2009; Hadmoko and Mauro, 2012; Jonkman et al., 2012; Lomnitz and Wisner, 2012; Schmuck, 2012). For instance, Galveston County, Texas adopted local emergency management and land use elements within a hazard mitigation plan before being affected by Hurricane Ike

2 These elements include: 1) increasing public awareness of the risks hazards pose; 2) assessing the adequacy of existing hazard mitigation measures (building codes, floodplain); 3) using these assessments as a means for citizens and local officials to create a vision for hazard resilience; 4) developing consensus on the need to take action (involving citizens and community intentions); 5) improving the likelihood that community stakeholders and residents will become aware of different hazard reduction approaches; 6) helping to guide the day-to-day decision making of local officials; 7) coordinating the action of various local government departments; and 8) providing the rational nexus between widely varying public interests that deal with safety from hazards.

Table 8.1 Literature on community-based disaster response planning by disaster type

Disaster types	Exposure and vulnerability	Disaster response planning			
		Generic development planning		Tourism planning	
Hydro-meteorological /climatological hazards					
Coastal storm	• Exposure of low lying coastal areas • Large population and megacities • Sea-level rise	• Structural mitigation (dikes, storm surge barriers, sea walls) • Non-structural mitigation (land use planning) • Ecosystem based disaster risk reduction (e.g., mangroves, coastal vegetables)	*Jonkman et al. (2012), Burby (1998)*	• Communication plans with disaster management [tourism industry]	*Higgins (2005), Johnston et al. (2007)*
Thunder storm and Tornado	• Severe thunderstorm beneficial and harmful impacts	• Land use planning and management (e.g., changing flash flood prone land from housing developments to parks or wetlands)	*Etkin et al. (2012), Burby (1998)*		
Flood	• Depth and velocity of water, speed of water-level rise, duration of the flood	• Structural (dikes, dam, levee, river bank, relocation, building elevation) and non-structural measures (zoning, land use) • New integrated flood defense approach with water resource management and pollution control	*Schmuck (2012), Burby (1998)*	• Community-wide early warning system [tourism industry and emergency service]	*Faulkner and Vikulov (2001)*
Drought	• Reduced crop productivity • Loss of wildlife habitat • Elevated livestock and wildlife mortality • Reduced availability of water resources	• Early warning system • Relationship between 'expert' information system and local system of community-based disaster mitigation and response • In agricultural communities, investment in soil and water conservation techniques (proactive drought risk reduction)	*Smucker (2012), Smucker and Wisner (2007)*		

Disaster types	Exposure and vulnerability	Disaster response planning			
		Generic development planning		Tourism planning	
Hydro-meteorological /climatological hazards					
Extreme Heat and Cold	• Vulnerable social status and networks, living and working environments, access to health service	• Community-based measures in preventing heat-and cold-related mortality • As a long term adaptation, promoting social networks (community-based disaster preparedness)	*McCormick (2012)*		
Wildfire	• Devastation as forests and other rural areas burned by wildfire	• Overarching legislation, dealing with all aspects of wildfires • A single fire/ land management organization responsible for wildfire management (e.g., fire-resistant site design) • Annual independent monitoring and public reporting on outcomes	*Underwood and Held (2012), Burby (1998)*	• Emergency management planning and agencies [location of tourism business]	*Hystad and Keller (2008)*
Geophysical hazards					
Landslide and other mass movement	• Reducing water quality • Lessening soil fertility and reducing stream and lake capacity • Physical and human loss	• Physical protection in facing landslides (e.g., covering soil cracks with ground coatings) • Identifying the potential causes of landslide (community-based activities) • Bioengineering solution (e.g., tree planting) Warning system	*Hadmoko and Mauro (2012), Burby (1998)*		
Earthquake	• Collapse of structures and secondary hazards such as fire, landslide and tsunami	• Performance-based engineering (e.g., improving building practices) • Hazard assessment • Building code • Earthquake early warning system • Community-based loss prevention	*Lomnitz and Wisner (2012), Burby (1998)*	• Purchasing earthquake insurance • Buildings and facilities reinforcement [tourism industry]	*Tsai and Chen (2010)*

Tsunami	• Vulnerable coastal communities	• Tsunami science • Risk assessment (communication network) • Coastal zone planning • The quality of building • Early warning system	*McAdoo (2012)*	• Communicating primary and solidarity messages to tourists and travel intermediaries [tourism industry] • Adopting new precautionary measures to protect the safety of hotel guests [tourism industry]	*Garcia et al. (2006), Carlsen (2006)*
Volcanic eruption	• Volcanic risk areas (benefits of living in an area-fertile soil, mineral wealth, geothermal energy, tourism potential	• Structural mitigation engineering solution to modify or confine the path of lahar • Non-structural mitigation (information on likely impacts and action, risk communication, monitoring and communicating network)	*Jenkins and Haynes (2012)*		

Note: Relevant literature identified in italics.

in 2008 (Bedient, 2012). The plan fostered greater resilience to natural hazards. (But see also Olonitua and Ibitoyo, 2014.)

Similar to these general disaster responses, other studies (e.g., Carlsen 2006; Faulkner and Vikulov 2001; Garcia et al. 2006; Higgins 2005; Hystad and Keller 2008; Johnston et al 2007) stressed that warnings targeting tourists and tourism-dependent businesses are imperative to reduce disaster event losses. As part of non-structural mitigation approaches, Tsai and Chen (2010) suggest that disaster management plans and insurance play a pivotal role in mitigating tourism–related losses following disaster events.

Disaster Resiliency, Community Responses, Planning, and Tourism

With an emphasis on disaster risk, resiliency refers to "the capacity of people and systems involved [returning] to a state prior to the occurrence of significant disturbances" (Garmestani et al., 2006: 535). If resiliency is improved, it will require less time to return to some form of normalcy or regular operation (Hamzah and Hampton, 2013). As a consequence, resiliency (another term for coping ability) is dependent on individuals and resources as well as on competencies to manage demands, challenges and changes. Likewise, resilient systems require flexibility and openness to change (Berkes and Ross, 2013; Schmidt and Garland, 2012). This approach assumes the fundamental recognition that future events are unexpected.

On the basis of community resilience to natural disasters, extensive literature (see Cutter et al., 2008; Ewing et al., 2010; Jha et al., 2013; Kapucu et al., 2013; Peacock et al., 2012; Wilkinson, 2012) summarizes resilience related variables to encompass ecological, institutional, social, economic, infrastructure, and community systems. As pointed out by Brody and Gunn (2013), ecological resiliency to flooding relies on a variety of physical characteristics such as floodplain area, soil permeability, wetlands acreage and loss, erosion rates, impervious surfaces, precipitation, and biodiversity characteristics. In addition, Cutter et al. (2008) observed that participation in hazard reduction programs, hazard mitigation plans, emergency services, zoning and building standards, emergency response plans, and continuity of operations plans can be included in institutional system resiliency.

Social and economic resilience, and community competence indicators involve demographic characteristics, social networks, community value cohesion, faith-based organizations, employment characteristics, property values, wealth generation, municipal fiscal elements, health and wellness, and quality of life (Aldrich, 2012; Berkes and Ross, 2013; Ersing and Kost, 2012; Nowell and Steelman, 2013; Zautra et al., 2009). In addition, as described by Cutter et al. (2008), infrastructure resilience factors include lifelines and critical infrastructure, transportation networks, residential housing stock and age, and commercial and manufacturing establishments.

Based on the theoretical foundation of the relationships between social and economic conditions and disaster resiliency, specific locally relevant tools and techniques for enhancing and strengthening community planning (associated with resiliency to natural disasters) were identified in the works of Beatley (2009) and Olshansky and Kartez (1998). As illustrated in Table 8.1, such policies included: 1) land use planning; 2) local infrastructure and public facilities; 3) taxation and financial incentives; 4) conservation and restoration of natural systems; and 5) building and structural resilience.

First, local land use planning as one of the non-structural flood mitigation approaches, allows communities to be more resilient to flooding (Gruntfest, 2000). This resiliency is enhanced by emphasizing cluster zoning (and other land use conservation measures) for future development, in accordance with the hazard mitigation strategy incorporating setbacks from natural features like dune systems, levees, flood walls, and wetlands (Beatley, 2009; Brody and Highfield, 2013; Jha et al., 2013). A second element of land use planning for community disaster resilience includes the design of local infrastructure and public facilities (Berkes et al., 2014). This local infrastructure element requires an alignment of land use decisions with public facilities and infrastructure investments determined by the community's carrying capacity, anticipated demands, and financial feasibility (Bedient, 2012). Likewise, in an effort to ensure more resilient facilities, communities need to assess municipal facilities to determine if structures could be strengthened or made more resistant to damage from such catastrophic events. This could include retrofitting for wind resistance, elevating buildings, or raising critical mechanical systems above anticipated flood levels.

When it comes to taxation and financial incentives, Beatley (2009) identified the fact that numerous communities have attempted to enhance resiliency by clustering small subdivisions on larger sized parcels. Through an expedited development review, waivers of permit fees, potentially short-term abatement of property taxes for new homeowners and developers in these areas, and capital improvements to infrastructure systems and neighborhood amenities, such financial incentives will decrease (Beatley, 2009).

The fourth community disaster resiliency policy relates to conservation and restoration of natural systems. In general, maintenance and development of natural vegetation and native land characteristics can improve air quality, provide shade, protect against erosion, lessen the impact of storm water, and serve as a wildlife habitat (Porter, 2000). Especially in coastal regions, restoring natural environmental characteristics allows for marshes and wetlands to provide a buffer from wave action. Likewise, in the recent work of "Greening in the Red Zone," Tidball and Krasny (2013) argued that creation of and access to green space confers resilience and recovery in social and ecological systems affected by violent conflict or disaster, and also promotes social health and wellbeing. This form of disaster-sensitive land conservation also serves to enhance tourism resiliency by adding to the supply of local outdoor recreation opportunities. This obvious connection results from a broader view of the tourism product where additional access to underlying natural amenities serves as the key motivator for leisure travel and subsequent local tourism-sensitive business receipts, as alluded to by Leiper (1979; 2008) and Marcouiller (2013).

A final policy is related to enhancements of structural resiliency, specifically natural hazard mitigation, to enhance drainage capacity and act to reduce recurrence of flooding. A recent example occurred in 2010 after Hurricane Ike. Galveston County, Texas undertook a system-wide cleaning of storm-related deposits with assistance from the Federal Emergency Management Agency (FEMA) (Bedient, 2012). Other infrastructure enhancements include providing curbs and gutters, paving alleyways, and regular street sweeping, which serve to lessen the amount of sand and yard debris entering sanitary systems (Beatley, 2009). Also, storm water retention systems and rain gardens (landscaped areas that hold water until it can be absorbed into the ground) can reduce storm sewer system demands during rain events. Recent studies (Orchiston, 2013) emphasize the critical role of disaster resilience planning in tourism-reliant communities by comparing the presence or absence of tourism business resilience in disaster prone areas. An example of this used resiliency planning relative to earthquake vulnerability in the Southern Alps of New Zealand. While resiliency planning for disasters does necessitate enhanced coordination, it is important to note that its underlying elements and their implementation are not new concepts. Indeed, resiliency planning is truly an extension of sound planning principles long espoused by the planning profession (c.f. Goodman and Freund, 1968). Pro-active approaches by community planners, locally elected public officials, and bureaucrats with respect to resiliency planning are central to its evolution and implementation. This is best summed up by Wagenaar and Wilkinson (2013: 15) with their suggestion of "enacting resilience" with respect to urban policy design discourse:

> As more and more public administrators turn to resilience as a frame for urban and environmental policy, the question of how to govern from a resilience perspective is raised.

Conceptual and Planning Relevant Models of Disaster Outcomes

Conceptual models of disaster vulnerability and resiliency are important frameworks upon which we can organize and discuss theoretical relationships. Indeed, these frameworks provide a critical basis upon which empirical models intent on describing the effects of disaster events on communities (vulnerability and risk), the planning response of communities to minimizing losses resulting from such events (resiliency), and integrative tourism and disaster planning over time, are formed.

A proposal for a Resiliency and Vulnerability Observatory Network (or RAVON), developed by Peacock et al. (2008: 4), outlines the imperative "to reduce the vulnerability associated with natural hazards and enhance the resiliency of

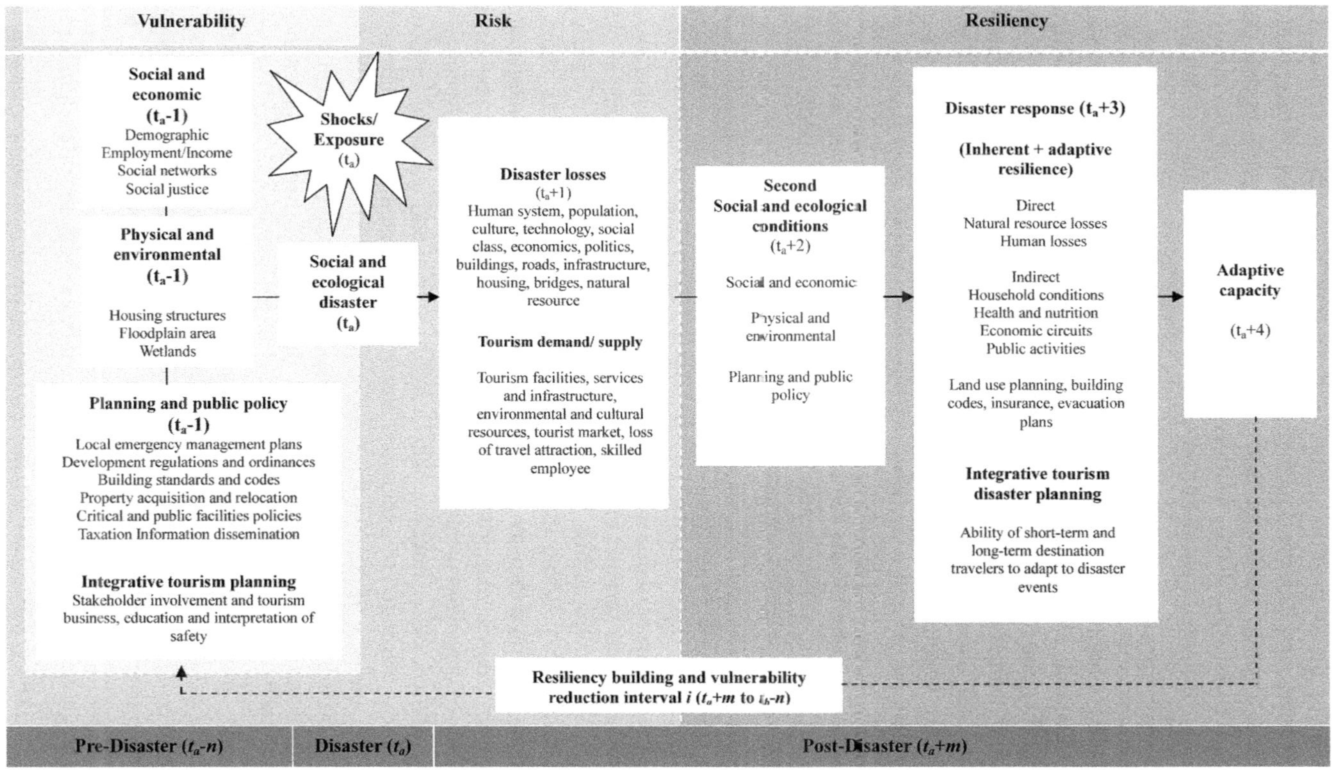

Figure 8.1 Integrative tourism disaster planning model

Note: Arrow indicates causal relationships, dotted arrow indicates feedback locp.

Source: Authors.

individuals and communities" through coordination among planning institutions. Others,[3] including the Disaster Resilience of Place (DROP) model (Cutter et al., 2008), the Pressure and Release (PAR) model (Blaikie et al., 1994; Wisner et al., 2004), and the Bogardi, Birkmann, and Cardona (BBC) conceptual framework (Bogardi and Birkmann, 2004; Cardona, 2004), mainly attempt to identify vulnerability or resiliency indicators derived from social and ecological elements in various disciplines.

Building upon this previous literature, we forward an integrative tourism disaster planning model in Figure 8.1. Note from this figure that we have employed spatial and temporal elements to reflect important conceptual insertions where both vulnerability and resiliency can affect event outcomes. Our notations t_a, t_a-*n*, and t_a+*m* reflect temporal elements where *t* denotes a point in time when a disaster event occurs, and subscript *a* represents the variety of locations (communities) affected by the event; *n* and *m* indicate time orders, with t_a-*n* reflecting a pre-disaster time period and t_a+*m* reflecting time post-disaster. Note further that we identify *i* as the interval between disaster events, which is equal to *n*+*m*. The interval between disaster events can serve as a key element involved in disaster planning effectiveness, as longer intervals can lead to complacency while shorter intervals can lead to increased urgency in resiliency planning.

Community Vulnerability Planning

The extent or forcefulness of disaster events makes up the central exogenous shock, while internal elements serve as exposure in the work of West and Lenze (1994). These authors and others outline factors that contribute to the extent of disaster losses (Raddatz, 2007). While the event itself serves as the exogenous shock, critical elements within a disaster-prone region can be thought of as endogenous exposures (or shocks) that exacerbate loss. These include internal conflict, political instability, lack of organized planning, and economic mismanagement (Acemoglu et al., 2003; Ahmed, 2003; Raddatz, 2007). In this spatial and temporal context, the extent of loss attributed to exogenous shock and endogenous exposure at t_a depends on the intensity of the disaster event, exposure (or vulnerability), and the prior disaster experience of the community. Disaster shocks have been defined as exceptional events operating outside "normal" development theory and practice (Baade et al., 2007).

Vulnerability to disaster, conceptualized as a potential for loss, refers to the likelihood of exposure to disaster damage (Etkin et al., 2004). In this context, vulnerabilities (t_a-1) during the first stage before a disaster period (t_a-*n*) include numerous characteristics that determine the ability of local institutions to absorb the shock of natural disaster events. Social and economic conditions of a community might have vulnerabilities that make it more likely to suffer significant losses relative to other adjacent, and therefore affected, communities. Conditions

3 These models are outlined in Table 8.2 and discussed further below.

that lead to increased exposure involve a multitude of items. These can include a lack of disaster preparedness and planning, relatively high unemployment, lower income levels, higher poverty rates, more isolated social networks, weak housing infrastructure, high levels of impervious surface, and low levels of preventive infrastructure (Pelling, 2003; Wisner, 2009). Based on these diverse factors, we can assess the degree of community vulnerability to disaster events.

In addition, community planning and public policy prior to a disaster event (t_a-n) further serve as important vulnerability criteria. These attributes that can serve as important exposures involve the extent to which the community has conducted local emergency management plans, land use planning, development regulations (zoning ordinances), building standards (building codes), property acquisition (building relocation), critical and public facilities, and taxation information dissemination (Beatley, 2009; Berkes et al., 2014; Burby, 1998; Olshansky and Kartez, 1998).

Vulnerability can also involve the extent to which tourism incidence affects community characteristics. Note from Figure 8.1 that we include a more integrative tourism planning process that accounts for past experience with local disaster events. Elements of this integrative tourism planning include stakeholder involvement of both long-term tourism migrants (second homeowners and retirees) and local tourism businesses accommodating short-term destination travelers. Further, education and safety with respect to disaster events can also serve as a source of vulnerability within disaster planning. In the process, it is necessary that the geographical and political characteristics of each community be considered with vulnerability factors that vary in accordance with differing community attributes.

As illustrated in Figure 8.1, the disaster losses element (t_a+1) as the first post-disaster stage (t_a+m) results from interactions among social and ecological systems (human systems, constructed systems, natural resource systems) and their many subsystems (population, culture, technology, social class, economics, politics, infrastructure, and tourism). Disaster losses change a community's social and ecological condition and can affect future exposure to subsequent disaster events. Unlike previous studies focused on vulnerability assessment composed of shocks, exposure, seasonality, trends, institutional structure, and processes (cf. Bogardi and Birkmann, 2004; Bohle, 2001; Chambers and Conway, 1992; Turner et al., 2003), the vulnerability framework outlined in Figure 8.1 addresses an integrated social and ecological sphere within a community-level planning and public policy context.

Community Resiliency Planning

As suggested by Rose (2007: 385), the resiliency of a community to a disaster event includes two properties labeled inherent resiliency and adaptive resiliency. Inherent resiliency is defined as "… the ordinary ability to deal with crises," while adaptive resiliency refers to "… the ability in crisis situation(s) to maintain function on the basis of ingenuity or extra effort." The conditions (t_a+2) as a second stage after a disaster (t_a+m) has been addressed in the existing literature (cf. Brody and Gunn, 2013; Cutter et al., 2008; Kapucu et al., 2013; Pelling, 2012; Wisner, 2009) to include social and economic systems (demographic, employment, income, social networks), physical and environmental systems (housing structure, floodplain area, wetlands, impervious surfaces), and community planning and public policy (disaster plans, development regulations, building standards, property acquisition, public facilities, taxation).

Especially when it comes to tourism disaster management and planning, Faulkner (2001) "… stresses coordinated team approaches, consultation and commitment in the disaster management process, but recognizes that natural disasters in particular are triggered by events over which victims have little or no control and that, therefore, their impacts are to some degree unavoidable. However, he [Faulkner] does suggest that their impacts can be moderated by good planning and management practices" (as cited in Cooper and Erfurt, 2007: 243). Within Faulkner's tourism disaster management framework, six elements of disaster management response involve precursors, mobilization, action, recovery, reconstruction and reassessment, and review. Within our discussion of Rose's two forms of resiliency, pre- "action" can be associated with "inherent resilience" and the remaining steps (from action to review) can be regarded as "adaptive resilience."

Further, as an alternative to the sustainable development paradigm, Lew (2014: 14) discusses the role of tourism within disaster planning and points out that "resilience planning has emerged … to provide new perspectives on community development and socio-ecological adjustments in a rapidly changing world." In the context of resilience to disaster damage, Rivera and Settembrino (2013) claimed that wealth enables a community to absorb and recover from disaster losses more quickly owing to insurance, social safety nets, social networks, and entitlement programs. Likewise, by integrating various factors of tourism planning within broader socio-cultural, economic, political, and environmental components, communities become more successful in managing disaster risk and mitigating losses that result from disaster events (Chhabra and Phillips, 2009).

Further along in the post-disaster resiliency planning phase, elements of adaptive resiliency become more evident. As a third stage (t_a+3) in the post-disaster (t_a+m) of Figure 8.1, new community conditions damaged by natural disasters will show a considerable number of disaster responses such as natural resource losses, household conditions, and community planning. For instance, examining communities with a more sizable economy before the disaster can be useful in determining whether or not they experience fewer disaster losses and require a shorter amount of time to recover from disasters. In the process of disaster response (t_a+4), communities with an adaptive capacity can mitigate or reduce disaster losses and shorten the period of disaster recovery. As a fifth stage (t_a+5) post-disaster, this process leads to vulnerability reduction and a new disaster preparedness and mitigation stage.

Integrative Tourism Disaster Planning Model

With respect to the tourism literature, disaster planning has become an increasingly interesting line of academic inquiry. We outline select tourism disaster work in Table 8.2. By integrating tourism within both disaster vulnerability and resiliency models (see Figure 8.1), we overcome the limitations presented by models myopically focused on vulnerability assessment. This builds from "The Onion Framework" as suggested by Bogardi and Birkmann (2004) and the Department for International Development's sustainable livelihood framework (DFID, 1999), which describes vulnerability as associated with different hazard impacts related to economic and social spheres. Also, both the Pressure and Release (PAR) model developed by Blaikie et al. (1994) and Wisner et al. (2004) and the ISDR framework for disaster reduction forwarded by the United Nations International Strategy for Disaster Reduction (UN/ISDR, 2004), attempted to assess the linkage between vulnerability and the development of a potential disaster. In the PAR model vulnerability was considered a process that includes increasing pressure and evolving opportunities. It categorized vulnerability factors into root causes, dynamic pressure, and unsafe conditions. Encompassing environmental, social, and economic spheres with risk assessment, the Bogardi, Birkmann, and Cardona (BBC) conceptual framework, depicted by Bogardi and Birkmann (2004) and Cardona (2004), addressed exposed and vulnerable elements and coping capacity as factors of vulnerability. With an emphasis on risk management, Cardona (2004) divided "risk" into two major categories—hard risk and soft risk. Whereas hard risk indicates potential damage impacts on physical infrastructure and environment, soft risk reflects potential socio-economic impacts on communities and organizations.

Furthermore, the Disaster Resilience of Place (DROP) model (Cutter et al., 2008) derived from the hazard-of-place model (Cutter, 1996; Cutter et al., 2000) designs the integration of vulnerability (antecedent condition) and resiliency assessment with an emphasis on varied factors associated with resiliency (ecological, social, economic, institutional, infrastructure, and community competence). This model includes disaster management cycles such as preparedness, mitigation, response, and recovery. By incorporating the relationship between tourism and resilience, the Scale, Change, and Resilience (SCR) model developed by Lew (2014) suggests four types of change in accordance with the degree of disturbance (from gradual to sudden) and scale of tourism actors (from private entrepreneurs to shared public interests). These change types include: 1) entrepreneur tourism slow change and 2) sudden shock; 3) community tourism slow change and 4) sudden shock. Further, this model encompasses four resilience indicators that include: 1) maintenance program; 2) training and diversification; 3) natural and cultural conservation; and 4) social support system.

As a critique, the SCR model fails to account for risk assessment, comparisons of alternative forms of disaster events, and the role of pro-active and integrative planning in disaster management. The integrative tourism disaster planning model

Table 8.2 Literature related specifically to vulnerability, resiliency, and risk models

Literature and referenced framework	Vulnerability		Resiliency		Risk	
	Factors	Characteristics	Factors	Characteristics	Factors	Characteristics
Bohle (2001): Conceptual framework for vulnerability analysis	• Exposure (the external side of vulnerability) • Coping (the internal side of vulnerability)	• Human ecology perspectives • Entitlement theory • Action theory approaches • Models of access to assets				
DFID (1999): The sustainable livelihood framework	• Shock • Trends • Seasonality	• Livelihood assets (Human, natural, financial, physical, social capitals) • Influence and access livelihood strategies				
UN/ISDR (2004): ISDR framework for disaster risk reduction		• As a tool and a precondition for effective risk assessment			• Vulnerability • Hazards	
Turner et al. (2003): Vulnerable framework	• Exposure • Sensitivity • Resilience	• Human and environmental influences outside the place • Global environmental change	• Adaptation			

Bogardi/Birkmann (2004): The Onion framework		• Different hazard impacts related to economic and social sphere • Potential losses and damages				
Blaikie et al. (1994) and Wisner et al. (2004): The Pressure and Release (PAR) model	• Root causes • Dynamic pressure • Unsafe conditions	• Vulnerability and the development of a potential disaster • A process involving increasing pressure and opportunity			• Risk = Hazard × Vulnerability	
Cardona (2004): Holistic approach to disaster risk assessment and management	• Exposure and physical susceptibility • Social and economic fragilities • Lack of resilience or ability to cope and recovering	• Control system (Risk management)			• Risk = (Hazard, Vulnerability)	• Hard risk: potential damage on physical infrastructure and environment • Soft risk: potential socio-economic impacts on communities and organizations
Bogardi/Birkmann (2004) and Cardona (2004): BBC conceptual framework	• Exposed and vulnerable elements • Coping capacity	• Environmental, social, and economic sphere			• Environmental • Social • Economic	
Cutter (1996), Cutter et al. (2000): Hazard-of-place model	• Exposure	• Integrating exposure and social vulnerability				

Literature and referenced framework	Vulnerability		Resiliency		Risk	
	Factors	Characteristics	Factors	Characteristics	Factors	Characteristics
Cutter et al. (2008): Disaster resilience of place (DROP) model	• Antecedent conditions (inherent vulnerability, inherent readiness)		• Ecological (e.g., wetland loss) • Social (e.g., social network) • Economic (e.g., employment) • Institutional (e.g., hazard mitigation plans) • Infrastructure (e.g., commercial establishments) • Community competence (e.g., quality of life)	• Disaster management cycles (preparedness, mitigation, response, and recovery) • Feedback loop		
Lew (2014): Scale, change and resilience (SCR) in tourism			• Maintenance programs • Training and diversification • Natural and cultural conservation • Social support system: social welfare and infrastructure	• Tourism and resilience contexts (e.g., community tourism-slow change and sudden shock, entrepreneur tourism-slow change and sudden shock)		

outlined in Figure 8.1 addresses comprehensive understandings about community vulnerability and resiliency. In line with integrated social and ecological spheres, this model encompasses risk assessment and compares hazard and disaster in accordance with disaster management cycles. Furthermore, this framework attempts to address the role of disaster management planning in reducing or mitigating tourism development losses, with an emphasis on tourism demand and supply. As described in Figure 8.1, based on the premise of pro-actively interacting with both supply and demand aspects associated with leisure-based migration (Hall, 2008; Marcouiller, 1997), integrative tourism disaster planning (including resiliency planning) addresses short and long-term travel patterns and reflects the ability of short-term and long-term destination travel to bounce back from natural disaster losses (loss of travel attractions, tourism facilities, services and infrastructure, environmental and cultural tourism resources, tourist market, information).

Conclusions

Natural disasters become crises when unambiguous failures of public and private decision making create outcomes that interrupt local activity. Whereas disaster vulnerability reflects the frequency and intensity of naturally occurring large-scale materialized risks, disaster resilience refers to the capacity of people and organizations to develop adaptive responses to perturbations that protect communities from potential loss. Together with the basic premise that disaster effects are fundamental social process responses that require pro-active planning, a conceptual model of disaster loss factors involves regional exposure, shock, and loss within the context of inherent social system vulnerability and resiliency. Specific research questions involve alternative levels of community development and social system status and their influence on disaster resilience.

We develop and discuss a conceptual and integrated model that inserts tourism planning as a key element within disaster planning. This takes on both planning and decision-making frameworks that act to simultaneously minimize vulnerability elements of a community while maximizing resiliency elements. This conceptual model then leads to a host of empirical applications that remain for further research. Based upon a review of the extant literature, three central theoretical hypotheses are proposed. First, disaster damage will have a negative association with social and economic development metrics; second, higher levels of community development will result in lower disaster-related losses, and; third, establishing better planning efforts, social capital, and social justice systems prior to an occurrence of natural disaster will lead to reduced disaster losses. This approach improves on previous work by incorporating tacit stages of development, social capital, social justice, and distributional elements that speak to social and economic inequity. Results from this review of extant knowledge suggest that disaster damage likely has an association with social and economic structures, and that engaged social capital, more equitable distributional characteristics, and local pro-active planning in-place before a disaster results in lower disaster losses (Pearce, 2003).

In line with this concept of vulnerability and resiliency to disaster and socio-cultural, economic, political, and environmental components in the context of an integrative tourism planning framework, we describe the role of disaster management planning in reducing or mitigating tourism development losses along with tourism demand and supply. In particular, this integrative tourism disaster planning approach can be useful in addressing the ability of short- and long-term destination travel to bounce back from various disaster events. In this respect, our work and subsequent empirical research responds to Ritchie's (2009, 226) effort to describe the role of planning in tourism crisis and disasters:

> Tourism crisis and disaster literature appears to be dominated by response and recovery at the expense of planning and prevention. Such research can assist tourism managers and policy-makers to develop effective policies and strategies.

References

Acemoglu, D, S. Johnson, J. Robinson, and Y. Thaicharoen (2003). Institutional causes, macro-economic symptoms, volatility, crisis and growth. *Journal of Monetary Economics* 50: 49–123.

Ahmed, S. (2003). Sources of macroeconomic fluctuations in Latin America and implications for choice of exchange rate regime. *Journal of Development Economics* 72: 181–202.

Aldrich, D.P. (2012). *Building Resilience: Social Capital in Post-disaster Recovery.* Chicago: The University of Chicago Press.

Angel, R.J., H. Bell, J. Beausoleil, and L. Lein (2012). *Community Lost: The State, Civil Society, and Displaced Survivors of Hurricane Katrina.* Cambridge: Cambridge University Press.

Baade, R.A., R. Baumann, and V. Matheson (2007). Estimating the economic impact of natural and social disasters, with an application to Hurricane Katrina. *Urban Studies* 44: 2061–2076.

Beatley, T. (2009). *Planning for Coastal Resilience: Best Practices for Calamitous Times.* Washington, DC: Island Press.

Bedient, P.B. (2012). *Lessons from Hurricane Ike.* College Station TX: Texas A&M University.

Berkes, F. and H. Ross (2013). Community resilience: Toward an integrated approach. *Society and Natural Resources* 26: 5–20.

Berkes, P.R., W. Lyles, and G. Smith (2014). Impacts of federal and state hazard mitigation policies on local land use policy. *Journal of Planning Education and Research* 34: 60–76.

Birkmann, J. (2006). Measuring vulnerability to promote disaster-resilient societies: Conceptual frameworks and definitions, in J. Birkmann (ed.) *Measuring Vulnerability to Natural Hazards: Towards Resilient Societies*. New York: United Nations University, 9–54.

Blaikie, P.T., T. Cannon, I. Davis, and B. Wisner (1994). *At Risk: Natural Hazards, People's Vulnerability, and Disasters*. London: Routledge.

Bogardi, J. and J. Birkmann (2004). Vulnerability assessment: The first step towards sustainable risk reduction, in D. Malzahn and T. Plapp (eds) *Disaster and Society: From Hazard Assessment to Risk Reduction*. Berlin: Logos Verlag, 75–82.

Bohle, H.G. (2001). "Vulnerability and criticality: Perspectives from social geography," IHDP Update 2/2001, *Newsletter of the International Human Dimensions Programme on Global Environmental Change*, 1–7.

Brody, S.D. and J.R. Gunn (2013). Examining environmental factors contributing to community resilience along the Gulf of Mexico coast, in N. Kapucu, C.V. Hawkins, and F.I. Rivera (eds) *Disaster Resiliency: Interdisciplinary Perspectives*. New York: Routledge, 160–177.

Brody, S.D. and W.E. Highfield (2013). Open space protection and flood mitigation: A national study. *Land Use Policy* 32: 89–95.

Brown-Jeffy, S. and S. Kroll-Smith (2009). Recovering inequality: Democracy, the market economy, and the 1906 San Francisco earthquake and fire, in E.C. Jones and A.D. Murphy (eds) *The Political Economy of Hazards and Disasters*. Lanham, MD: Altamira Press, 83–103.

Buckle, P. (2006). Assessing social resilience, in D. Paton and D. Johnston (eds) *Disaster Resilience: An Integrated Approach*. Springfield, IL: Charles C Thomas, 88–104.

Burby, R.J. (1998). *Cooperating with Nature: Confronting Natural Hazards with Land-Use Planning for Sustainable Communities*. Washington, DC: Joseph Henry Press.

Burby, R.J. (2005). Have state comprehensive planning mandates reduced insured losses from natural disasters? *Natural Hazards Review* 6: 67–81.

Burby, R.J. and F. Wagner (1996). Protecting tourists from death and injury in coastal storms. *Disasters* 20: 49–60.

Campiranon, K. and N. Scott (2007). Factors influencing crisis management in tourism destinations, in E. Laws, B. Prideaux, and K. Chon (eds) *Crisis Management in Tourism*. Wallingford, UK: CAB International, 142–156.

Cardona, O.D. (2004). The need for rethinking the concepts of vulnerability and risk from a holistic perspective: A necessary review and criticism for effective risk management, in G. Bankoff and G. Freks (eds) *Mapping Vulnerability: Disasters, Development and People*. London: Earthscan, 37–51.

Carlsen, J. (2006). Post-tsunami tourism strategies for the Maldives. *Tourism Review International* 10: 69–79.

Chambers, R. and G. Conway (1992). "Sustainable rural livelihoods: Practical concepts for the 21st century." IDS Discussion Paper 296. Brighton: Institute of Development Studies.

Chhabra, D. and R. Phillips (2009). Tourism-based development, in R. Phillips and R.H. Pittman (eds) *An Introduction to Community Development*. London: Routledge, 236–248.

Chi, G. and D.W. Marcouiller (2012). Recreational homes and migration to remote amenity-rich areas. *Journal of Regional Analysis and Policy* 42: 47–60.

Cooper, M. and P. Erfurt (2007). Tsunamis, earthquakes, volcanism and other problems: Disasters, responses and Japanese tourism, in E. Laws, B. Prideaux, and K. Chon (eds) *Crisis Management in Tourism*. Wallingford, UK: CAB International, 234–251.

Cutter, S.L. (1996) Vulnerability to environmental hazards. *Progress in Human Geography* 20: 529–539.

Cutter, S.L. and C.T. Emrich (2006). Moral hazard, social catastrophe: The changing face of vulnerability along the hurricane coasts. *Annals of the American Academy of Political and Social Science* 604: 102–112.

Cutter, S.L., J.T., Mitchell, and M.S. Scott (2000). Revealing the vulnerability of people and places: A Case study of Georgetown County, South Carolina. *Annals of the Association of American Geographers* 90: 713–737.

Cutter, S.L., L. Barnes, M. Berry, et al. (2008). A place-based model for understanding community resilience to natural disasters. *Global Environmental Change* 18: 598–606.

Department for International Development (DFID) (1999). *Sustainable Livelihood Guidance Sheets*. London: DFID.

Ersing, R.L. and K.A. Kost (2012). *Surviving Disaster: The Role of Social Networks*. Chicago, IL: Lyceum Books, Inc.

Etkin, D., E. Haque, L. Bellisario, and I. Burton (2004). *An Assessment of Natural Hazards and Disasters in Canada* (The Canadian Natural Hazards Assessment Project; Public Safety and Emergency Preparedness). Ottawa: Environment Canada.

Etkin, D., K. Higuchi, and G. Platsis (2012). Thunderstorm and tornado, in B. Wisner, J.C. Gaillard, and I. Kelman (eds) *The Routledge Handbook of Hazards and Disaster Risk Reduction*. London: Routledge, 232–243.

Ewing, L., R.E. Flick, and C.E. Synolakis (2010). A review of coastal community vulnerabilities toward resilience benefits from disaster reduction measures. *Environmental Hazards* 9: 222–232.

Faulkner, B. and S. Vikulov. (2001). Katherine, washed out one day, back on track the next: A post-mortem of a tourism disaster. *Tourism Management* 22: 331–344.

Faulkner, W. (2001). Towards a framework for tourism disaster management. *Tourism Management* 22: 135–147.

Flora, C.B., J.L. Flora, and S. Fey (2003). *Rural Communities: Legacy and Change*. Boulder, CO: Westview Press.

Garcia, R., S.S. Lau, K.W. Chau, et al. (2006). Sustainable resorts: Learning from the 2004 tsunami. *Disaster Prevention and Management* 15: 429–447.

Garmestani, A., C.R. Allen, J.D. Mittelstaedt, et al. (2006). Firm size, diversity, functional richness, and resilience. *Environment and Development Economics* 11: 533–551.

Goodman, W.I. and E.C. Freund (eds) (1968). *Principles and Practice of Urban Planning.* Washington, DC: International City Managers' Association.

Green, G.P. and A. Haines (2012). *Asset Building & Community Development*. Los Angeles, CA: Sage.

Gruntfest, E. (2000). Nonstructural mitigation of flood hazards, in E.E. Wohl (ed.) *Inland Flood Hazards: Human, Riparian, and Aquatic Communities*. Cambridge: Cambridge University Press, 394–410.

Hadmoko, D.S. and S.E. Mauro (2012). Landslide and other mass movements, in B. Wisner, J.C. Gaillard, and I. Kelman (eds) *The Routledge Handbook of Hazards and Disaster Risk Reduction*. London: Routledge.

Hall, C.M. (2008). *Tourism Planning: Policies, Processes, and Relationships* (2nd edition). New York: Pearson Prentice Hall.

Hall, C.M. and D.K. Müller (2004). *Tourism, Mobility and Second Homes: Between Elite Landscape and Common Ground.* Tonawanda, NY: Channel View Publications.

Hamzah, A. and M.P. Hampton (2013). Resilience and non-linear change in island tourism. *Tourism Geographies* 15: 43–67.

Higgins, B.A. (2005). The storms of summer: Lessons learned in the aftermath of the hurricanes of '04. *Cornell Hospitality Quarterly* 46: 40–46.

Holcombe, E. (2010). Understanding community-based disaster response: Houston's religious congregations and Hurricane Katrina relief efforts, in D.L. Brunsma, D. Overfelt, and J.S. Picou (eds) *The Sociology of Katrina: Perspectives on a Modern Catastrophe*. Lanham, MD: Rowman & Littlefield Publishers, 119–132.

Hystad, P.W. and P.C. Keller (2008). Towards a destination tourism disaster management framework: Long-term lessons from a forest fire disaster. *Tourism Management* 29: 151–162.

Ivars Baidal, J.A. (2004). Regional tourism planning in Spain: Evolution and perspectives. *Annals of Tourism Research* 31: 313–333.

Jenkins, S. and K. Haynes (2012). Volcanic eruption, in B. Wisner, J.C. Gaillard, and I. Kelman (eds) *The Routledge Handbook of Hazards and Disaster Risk Reduction*. London: Routledge, 334–346.

Jha, A.K., T.W. Miner, and Z. Stanton-Geddes (2013). *Building Urban Resilience: Principles, Tools, and Practice*. Washington, DC: The World Bank.

Johnston, D., J. Becker, C. Gregg, et al. (2007). Developing warning and disaster response capacity in the tourism sector in coastal Washington, USA. *Disaster Prevention and Management* 16: 210–216.

Jonkman, S.N., H. Gerritsen, and M. Marchand (2012). Coastal storm, in B. Wisner, J.C. Gaillard, and I. Kelman (eds) *The Routledge Handbook of Hazards and Disaster Risk Reduction*. London: Routledge, 220–231.

Kapucu, N., C.V. Hawkins, and F. Rivera (2013). Disaster resiliency: Interdisciplinary perspectives, in N. Kapucu, C.V. Hawkins, and F.I. Rivera (eds) *Disaster Resiliency: Interdisciplinary Perspectives*. New York: Routledge, 1–14.

Kasperson, R.E. and J.X. Kasperson (2005). Climate change, vulnerability and social justice, in J.X. Kasperson and R.E. Kasperson (eds) *The Social Contours of Risk: Publics, Risk Communication and the Social Amplification of Risk* (Volume 1). London: Earthscan, 301–321.

Kelman, I., R. Spence, J. Palmer, et al. (2008). Tourists and disasters: Lessons from the 26 December 2004 tsunamis. *Journal of Coastal Conservation* 12: 105–113.

Leiper, N. (1979). The framework of tourism: Towards a definition of tourism, tourist, and the tourism industry. *Annals of Tourism Research* 6: 390–407.

Leiper, N. (2008). Partial industrialization in tourism: A new model. *Current Issues in Tourism* 11: 205–235.

Lew, A.A. (2014). Scale, change and resilience in community tourism planning. *Tourism Geographies* 16: 14–22.

Lomnitz, C. and B. Wisner (2012). Earthquake, in B. Wisner, J.C. Gaillard, and I. Kelman (eds) *The Routledge Handbook of Hazards and Disaster Risk Reduction*. London: Routledge, 310–321.

Lyles, L.W., P. Berke, and G. Smith (2014). Do planners matter? Examining factors driving incorporation of land use approaches into hazard mitigation. *Journal of Environmental Planning and Management* 57: 792–811.

Marcouiller, D.W. (1997). Toward integrative tourism planning in rural America. *Journal of Planning Literature* 11: 337–357.

Marcouiller, D.W. (2013). The rural development attributes of tourism, in G.P. Green (ed.) *The Handbook of Rural Development*. Cheltenham UK: Edward Elgar, 158–178.

McAdoo, B.G. (2012). Tsunami, in B. Wisner, J.C. Gaillard, and I. Kelman (eds) *The Routledge Handbook of Hazards and Disaster Risk Reduction*. London: Routledge, 322–333.

McCormick, S. (2012). Extreme heat and cold, in B. Wisner, J.C. Gaillard, and I. Kelman (eds) *The Routledge Handbook of Hazards and Disaster Risk Reduction*. London: Routledge, 269–280.

Miller, D.S. and J.D. Rivera (2007). Setting the stage: Roots of social inequity and the human tragedy of Hurricane Katrina, in K.A. Bates and R.S. Swan (eds) *Through the Eye of Katrina: Social Justice in the United States*. Durham, NC: Carolina Academic Press, 15–35.

Miller, J.L. (2012). *Psychosocial Capacity Building in Response to Disasters*. New York: Columbia University Press.

Nelson, A.C. and S.P. French (2002). Plan quality and mitigating damage from natural disasters: A case study of the Northridge earthquake with planning policy considerations. *Journal of the American Planning Association* 68: 194–207.

Nowell, B. and T. Steelman (2013). The role of responder networks in promoting community resilience: Toward a measurement framework of network capacity, in N. Kapucu, C.V. Hawkins, and F.I. Rivera (eds) *Disaster Resiliency: Interdisciplinary Perspectives*. New York: Routledge, 232–257.

Olonilua, O. and O. Ibitayo (2014). An In-Depth Analysis of the Houston–Galveston Area Council Regional Hazard Mitigation Plan under the Disaster Mitigation Act of 2000 (DMA2K). *Risk, Hazards & Crisis in Public Policy* 5: 316–341.

Olshansky, R.B. and J.D. Kartez (1998). Managing land use to build resilience, in R.J. Burby (ed.) *Cooperating with Nature: Confronting Natural Hazards with Land-Use Planning for Sustainable Communities*. Washington, DC: Joseph Henry Press.

Orchiston, C. (2013). Tourism business preparedness, resilience and disaster planning in a region of high seismic risk: The case of the Southern Alps, New Zealand. *Current Issues in Tourism* 16: 477–494.

Patterson, J. (2013). Equity in disasters: Civil and human rights challenges in the context of emergency events, in D.R. Gilmore and D.M. Standaert (eds) *Building Community Resilience Post-Disaster: A Guide for Affordable Housing and Community Economic Development Practitioners*. Chicago, Illinois: American Bar Association, 91–107.

Peacock, W.G., H. Kunreuther, W.H. Hooke, et al. (2008). *Toward a Resiliency and Vulnerability Observatory Network: RAVON*. HRRC reports: 08-02R. archone.tamu.edu/hrrc/publications/researchreports/RAVON.

Peacock, W.G., S.V. Zandt, D. Henry, et al. (2012). Using Social Vulnerability Mapping to Enhance Coastal Community Resiliency in Texas, in P.B. Bedient (ed.) *Lessons from Hurricane Ike*. College Station, TX: Texas A&M University Press, 66–81.

Pearce, L. (2003). Disaster management and community planning, and public participation: How to achieve sustainable hazard mitigation. *Natural Hazards* 28: 211–228.

Pelling, M. (2003). *The Vulnerability of Cities: Natural Disasters and Social Resilience*. London: Earthscan.

Pelling, M. (2012). Hazards, risk and urbanization, in B. Wisner, J.C. Gaillard, and I. Kelman (eds) *The Routledge Handbook of Hazards and Disaster Risk Reduction*. London: Routledge, 145–155.

Pezzullo, P. (2009). Tourists and disasters: Rebuilding, remembering, and responsibility in New Orleans. *Tourist Studies* 9: 23–41.

Phillips, B. and B. Morrow (2007). Social science research needs: Focus on vulnerable populations, forecasting, and warnings. *Natural Hazards Review* 8: 61–68.

Porter, D. (2000). *The Practice of Sustainable Development*. Washington, DC: Urban Land Institute.

Raddatz, C. (2007). Are external shocks responsible for the instability of output in low-income countries? *Journal of Development Economics* 84: 155–187.

Ritchie, B.W. (2009). *Crisis and Disaster Management for Tourism*. Bristol, UK: Channel View Publications.

Rivera, F.I. and M.R. Settembrino (2013). Sociological insights on the role of social capital in disaster resilience, in C.V. Hawkins, and F.I. Rivera (eds) *Disaster Resiliency: Interdisciplinary Perspectives*. New York: Routledge, 48–60.

Roca, Z. (2013). *Second Home Tourism in Europe: Lifestyle Issues and Policy Responses*. Burlington, VT: Ashgate.

Rodriguez, H. and C. Russell (2006). Understanding disasters: Vulnerability, sustainable development, and resiliency, in J. Blau and K.E. Smith (eds) *Public Sociologies Reader*. Lanham, MD: Rowman & Littlefield, 193–211.

Rose, A. (2007). Economic resilience to natural and man-made disasters: Multidisciplinary origins and contextual dimensions. *Environmental Hazards* 7: 383–398.

Schewe, R.L., D.R. Field, D.J. Frosch, et al. (2012). *Condos in the Woods: The Growth of Seasonal and Retirement Homes in Northern Wisconsin*. Madison, WI: University of Wisconsin Press.

Schmidt, D.H. and K.A. Garland (2012). Bone dry in Texas: Resilience to drought on the Upper Texas Gulf Coast. *Journal of Planning Literature* 27: 434–445.

Schmuck, H. (2012). Flood, in B. Wisner, J.C. Gaillard, and I. Kelman (eds) *The Routledge Handbook of Hazards and Disaster Risk Reduction*. London: Routledge, 244–256.

Schwab, A.K., K. Eschelbach, and D.J. Brower. (2007) *Hazard Mitigation and Preparedness: Building Resilient Communities*. New York: John Wiley & Sons.

Smith, G. (2011). *Planning for Post-disaster Recovery: A Review of the United States Disaster Assistance Framework*. Fairfax, VA: The Public Entity Risk Institute.

Smucker, T. and B. Wisner (2007). Changing household responses to drought in Tharaka, Kenya: Persistence, change, and challenge. *Disasters* 32: 190–215.

Smucker, T.A. (2012). Drought, in B. Wisner, J.C. Gaillard, and I. Kelman (eds) *The Routledge Handbook of Hazards and Disaster Risk Reduction*. London: Routledge, 257–268.

Stone, G.W., J.M. Grymes, J.R. Dingler, and D.A. Pepper (1997). Overview and significance of hurricanes on the Louisiana coast, USA. *Journal of Coastal Research* 13: 656–689.

Swan, R.S. and K.A. Bates (2007). Social justice in the face of the storm: When natural disasters become social disasters, in K.A. Bates and R.S. Swan (eds) *Through the Eye of Katrina: Social Justice in the United States*. Durham, NC: North Carolina Academic Press, 3–12.

Tidball, K.G. and M.E. Krasny (2013). *Greening in the Red Zone: Disaster, Resilience, and Community Greening*. Dordrecht: Springer.
Tierney, K. (2014). *The Social Roots of Risk: Producing Disasters, Promoting Resilience*. Stanford, CA: Stanford Business Books.
Tsai, C.H. and C.W. Chen (2010). An earthquake disaster management mechanism based on risk assessment information for the tourism industry: A case study from the Island of Taiwan. *Tourism Management* 31: 470–481.
Turner, B.L., R.E. Kasperson, P.A. Matson, et al. (2003). A framework for vulnerability analysis in sustainability science. *Proceedings of the National Academy of Sciences* 100: 8074–8079.
UN/ISDR (International Strategy for Disaster Reduction) (2004). *Living with Risk: A Global Review of Disaster Reduction Initiatives*. Geneva: UN Publications.
Underwood, R. and A. Held (2012). Wildfire, in B. Wisner, J.C. Gaillard, and I. Kelman (eds) *The Routledge Handbook of Hazards and Disaster Risk Reduction*. London: Routledge, 281–293.
Vollmer, H. (2013). *The Sociology of Disruption, Disaster and Social Change: Punctuated Cooperation*. Cambridge: Cambridge University Press.
Wagenaar, H. and C. Wilkinson (2013). Enacting resilience: A performative account of governing for urban resilience. *Urban Studies* 52: 1265-1284.
West, C. and D. Lenze (1994). Modeling the regional impact of natural disaster and recovery: A general framework and an Application to Hurricane Andrew. *International Regional Science Review* 17: 121–150.
Wilkinson, C. (2012). Social-ecological resilience: Insights and issues for planning theory. *Planning Theory* 11: 148–169.
Wisner, B. (2009). "SHINK & swim: Exploring the link between capital (social, human, institutional, natural), disaster, and disaster risk reduction." Background paper for EDRR project, GFDRR. Washington, DC: World Bank.
Wisner, B., P. Blaikie, T. Cannon, and I. Davis (2004). *At Risk: Natural Hazards, People's Vulnerability and Disasters*. London: Routledge.
Woosnam, K.M. and H. Kim. (2014). Hurricane impacts on Southeastern United States coastal national park visitation. *Tourism Geographies* 16: 364-381.
Zautra, A., J. Hall, and K. Murray (2009). Community development and community resilience: An integrative approach. *Community Development* 39: 130–147.

Chapter 9
Computing the Aggregate Cost of a Disaster: National and Regional Issues

Charles Steindel

In the wake of a disaster such as Superstorm Sandy, there is widespread interest in its overall dollar "cost." Much of the attention to such numbers might be dismissed as mere curiosity (is the current disaster "officially" bigger than one in the past?), but there are parties to whom timely information of this type is important. Most notably, macroeconomic analysts, such as those at the Federal Reserve, need to discern the impact of the event on national data in order to segregate disaster impacts from the cyclical and secular trends that are the primary inputs in policy-making. For example, virtually every page of the transcript of the Federal Open Market Committee meeting on September 20, 2005 has some discussion of the potential impact of Hurricane Katrina (http://www.federalreserve.gov/monetarypolicy/files/FOMC20050920meeting.pdf). While much of the talk revolved around the impact of the storm on energy markets, there were also references to the extent of physical damage and output disruptions. Indeed, the "Greenbook"—the summary of the staff's outlook—began with several pages discussing these issues.

There is also a need for timely estimates of the cost of a disaster on a smaller scale than nationwide. State and local officials are concerned, since they need to compute the effect of the event on government finances—both on the revenue side (how much, and how long, will revenues be depressed?) as well as on the expenditure side. Moreover, timely estimates of the state or regional cost of a disaster are used in framing requests for federal aid.

This chapter explores some of the issues involved in defining the aggregate cost of a disaster and the current availability of reliable estimates, zeros in on the recent experience of Sandy as an illustration of how such estimates are produced and used, and makes some suggestions for improvements to the method of acquisition and reporting of numbers on aggregate disaster costs.

Conceptualizing Aggregate Disaster Costs

An initial issue is defining what is meant by a disaster and its costs. Much of the literature on evaluating the aggregate cost of disasters appears to be oriented around profound events, such as the Haitian earthquake, which involve severe disruptions to a nation's infrastructure (Pelling, Özerdem, and Barakat [2002] and

Barrito [2008] discuss the impact of profound disasters on national economies). The more typical case in the US is an event that renders unavailable, or destroyed, a noticeable portion of the production inputs of a fairly well specified area. Such events, though they may inflict considerable dollar costs, tend to be localized in their effects. (Listings of major US events and estimates of physical damages resulting from them may be found at http://www.ncdc.noaa.gov/billions/events and http://webra.cas.sc.edu/hvri/products/sheldus.aspx.)

In considering the aggregate impact of this type of localized disaster it is useful to distinguish between its effect on the wealth of the people of the affected region and its effect on the region's output.

We will first address the wealth effect. The wealth of the residents of the area may be expressed as:

(1) $$\int (w_t LR_t + TRANS_t) e^{-rt} dt + NHW_0$$

where

w_t = hourly wage received by workers who were residents of the affected area at the time of the disaster;

LR_t = aggregate hours worked by workers who were residents of the affected area at the time of the disaster;

$TRANS_t$= net transfers to persons who were residents of the affected area at the time of the disaster;

NHW_0 = nonhuman wealth of residents of the affected area at the time of the disaster, and;

R = rate of discount.

A disaster will alter the time path of the hours worked by residents, surely reducing it in the short run. Wage rates will also likely be altered. In addition, residents' physical wealth will be reduced to the extent they own assets damaged, destroyed, or whose market value has otherwise fallen as a result of the disaster (naturally, nonresident property owners will also suffer losses). A logical measure of the cost of a disaster to the area's residents is the present discounted value of the external transfer inflow necessary to restore aggregate wealth of the directly affected population to its initial value.

The impact of a disaster on the output of an area is conceptually different than its effect on the wealth of the resident population. The value of the output of a region is:

(2) $p_{t*} F_t(L_t, K_t, E_t, M_t) + \mu_t,$

where

F_t = production technology used in the region in time t (for simplicity, the region's output is assumed to consist of only one product);

p_t = price of the output produced in the region in time t;

L_t, K_t, E_t, M_t = labour, capital, energy, and other intermediate products employed in the region, and;

μ_t = an exogenous shock to the production process in time t.

A disaster will likely alter the time paths of the inputs to production in the region, may be associated with a negative exogenous shock, and may result in changes to the production technology used in the region (a change in the production function, not merely a shift in the factors). The demand for output of the region may also change, affecting its price. In this setup spillovers from disasters in one area to production in another area are ignored. However, some recent episodes, such as 9/11 and Hurricane Katrina, saw notable spillovers of this type. Blomberg and Hess (2009) estimate the spillovers to the nation from 9/11; Greenberg et al. (2013) discuss the potential regional (New York and New Jersey) effects of a hypothesized New Jersey railroad disaster.

The impact of a disaster on a region's output could be viewed as the change in the present discounted value of production over some *fixed period*, including and following the disaster. It is possible that over an extended period of time the disaster may have a positive impact on output, and taking a sufficiently long horizon one might even conclude that there is little or no change in the present value of production. However, comparing scenarios with and without a disaster over very long horizons would be a very problematic undertaking.

There is an obvious asymmetry between the wealth and output costs: wealth costs in principle should reflect changes in income over an infinite (or very lengthy) horizon, while output effects are usually computed over a fairly short period. For example, a common question raised after a disaster is when output in the region directly affected (say the New Jersey shore, in the case of Sandy) will return to "normal." Presumably, normality means a return to either the pre-disaster level or the path of output.

Why should there be interest in the short-run fluctuations in regional output (and employment) associated with a disaster? One reason is that estimates of the loss of output—its total and its timing—from a disaster may be useful to states projecting the potential revenue impact of the event. Estimates of the near-term output and revenue impact of a disaster are not only important for normal budgeting purposes, but are also requested by ratings agencies and investors in municipal securities.

Such concerns may often be traced to fairly strong "balanced budget" constraints in many states (often meaning constitutional prohibitions against borrowing to cover revenue shortfalls), so estimates of the short-term revenue effect of a disaster may be very important. New Jersey has this type of tight borrowing constraint, and Sandy hit in the middle of fiscal year 2013, during the period in which the fiscal year 2014 budget was being prepared.

In the wake of a disaster the distinction between types of costs can be confused. Following a disaster, there is a marked focus on the extent of physical damage (particularly in real terms)—how many homes and cars were destroyed, how many businesses suffered serious losses, and so on. Estimates of physical damage and its cost in dollars are important since they, along with private insurance reimbursements, will be major factors in determining the amount of net transfers required to make whole the population of the affected area. Of course, it should be acknowledged that the physical damage to fixed property in the area of disasters need not equal the wealth loss of residents. Most notably, a large portion of the wealth of residents will be the present value of their wages (also, as noted above, property in the area may be owned by nonresidents).

Physical damage and other losses of wealth resulting from a disaster are reasonably straightforward to conceptualize and measure (in practice, of course, the actual estimation may be complex). Determining the output effects of a disaster raises rather complex questions. The answers depend on the specific time interval and region being asked about. If the questions deal with the region directly affected, any or all of the elements entering into the determination of nominal output in equation (2) may be affected by the disaster:

1. The demand for production in the region may shift. In the immediate aftereffect of Sandy there were widespread cancellations of 2013 reservations to stay at the New Jersey shore, which may be interpreted as a downward shift in the demand curve for New Jersey shore output, working to reduce its price.
2. The production technology used in an area may change. Following a disaster, the affected region may produce different goods and services or may produce the same items using different techniques.
3. Various inputs may be more or less available. For instance, liquid fuels and electricity were in short supply in New Jersey and New York just after Sandy, and transportation and other problems inhibited workers' ability to travel to their jobs. Also, more permanent displacement of labor and customers will affect the time path of output. (Xiao and Van Sandt [2012] provide evidence from the aftermath of Hurricane Ike on the interrelationship between household rebuilding following a natural disaster and business recovery.)
4. The disaster may be associated with a sequence of shocks to the production process—possibly initially negative, then positive. Such shocks should be distinguished from shifts in the availability of inputs.

Availability of Estimates of Disaster Costs

Estimates of physical damage caused by disasters are relatively assessable. Often, such estimates are expressed in physical, not dollar terms (so many homes damaged or destroyed, for instance). However, in the US, the Commerce Department's Bureau of Economic Analysis (BEA), using data from FEMA and other sources such as private insurers, routinely compiles estimates of the dollar cost of the damage inflicted by major disasters. Unlike many of the numbers disseminated in the wake of a disaster, these figures are produced in a systematic accounting framework. Nonetheless, the BEA's estimates have some gaps limiting their usefulness:

- They provide information at the national level only. For some events, such as 2008's Hurricane Ike (which was essentially confined to Texas) the national aggregate is sufficient. For many others, such as Hurricane Katrina and Superstorm Sandy, the lack of breakdown at the state level becomes an obstacle to understanding the specific regional impacts of the disaster. Ironically, the BEA's older treatment of disasters, in which disaster damage was viewed as an increase in capital consumption and charged to income from current production, allowed for a better analysis of regional effects, since uninsured private damage would directly depress components of quarterly state personal income (especially rental and proprietors' income series). In the current treatment, destruction resulting from a large disaster is not included in the capital consumption series and is not charged against current national or state incomes.
- The BEA's damage estimates, in line with their usual accounting for capital valuation, are of the decline in the reproduction cost valuation of depreciated assets. That loss may or may not equal the loss in the market value of (nonhuman) wealth due to a disaster. The BEA's estimates also do not take into account the installation costs of replacement assets. This is of particular concern for damaged public infrastructure: the cost of repairing washed-out roads and bridges may well exceed the value of the loss of the depreciated assets.
- The list of assets included in the accounting is fairly narrow. The BEA does not estimate the loss of business inventories from a disaster. These losses would be included in the usual reports on business inventories, and would thus subtract from the Gross Domestic Product (GDP) in the quarter in which they occur (see the Appendix for a discussion of the potential effect of Sandy destruction on business inventory investment, especially in automobiles). Also, the BEA does not estimate the amount of disaster loss to personal property, including automobiles, furniture, and appliances.

Existing US official economic data systems do not systematically account for the effects of disasters on national or regional output, especially in a timely fashion (though the Federal Reserve does typically provide assessments of how events of this type may have affected national monthly industrial production figures). One shortcoming is the lack of high frequency subnational output data. Even at the national level, estimates of output at any frequency higher than monthly are problematic except for weekly unemployment claims and some measures of physical industry production (auto assemblies and the like). State-level data are highly limited at the monthly level, and are often issued only after a significant lag. Most notably, estimates of state gross product are only available at an annual frequency, and initial estimates for a year are not released until the middle of the following year. Clearly, data availability problems are accentuated at the sub-state level. We can observe how payrolls and unemployment claims change in an area in the wake of a disaster, but there is little else in the way of reliable information related to output. Ultimately, though, to compute the effect of a disaster on output we need to compute the counterfactual: what would output have been without the disaster?

A major obstacle to computing a counterfactual, thus fine-tuning the effect of a disaster on regional output, is the lack of specific information on the production technology of an area, which is necessary to model the region's economy, including its interactions with other areas.[1] In the absence of such information analysts must make their own, potentially arbitrary, assumptions. For example, Guimaraes, Hefner, and Woodward (1993) use an econometric model of the state to simulate how Hurricane Hugo affected the output of South Carolina. The model was estimated using pre-Hugo data—in effect, freezing the state's production technology and demand prior to the storm—then inferring the storm's effects from out-of-sample errors. Bram, Orr and Rapaport (2002) estimate the effect of the September 2011 attacks on New York City employment from out-of-sample errors of autoregressions; the impact on aggregate wages (which would be a major component of the impact on output) was then readily inferred. It is clear that both procedures are heavily based on the reliability of pre-disaster model estimates, as well as on other assumptions (Guimaraes, Hefner, and Woodward assumed that the national economic environment was not affected by Hugo, while Bram, Orr, and Rapaport produced estimates that assumed the US baseline also changed after 9/11). Ultimately, one's acceptance of the value of such estimates depends on one's assessment of the value of modeling assumptions used. Greenberg, Lahr and Mantell (2007) review the types of models that may be used to compute the time series of regional economic impacts of disasters.

The dependence on modeling assumptions also applies to estimates made with the use of FEMA's tools. FEMA's Hazards United States (HAZUS) models allow users to compute full-blown regional costs (both physical damage and output loss)

1 Cochrane (1997) describes some of the channels though which a disaster affects regional and national output. The literature on the economic effects of disasters emphasizes the need for formal modeling to better calibrate their regional impacts (e.g. Greenberg, Lahr, and Mantell, 2007).

of various types of disasters but, like the examples noted above, these estimates will clearly be heavily dependent on highly specific assumptions about the effects of disasters on specific elements of a region's capital and infrastructure, as well as on details of the input-output structure of the affected area's economy and how it may evolve over time.

In the end, estimates of output costs derived from models such as these may be useful in thinking about long-term planning for a region following a disaster, but are not likely to be available to provide especially informative timely estimates of the effect of a particular event. Estimates based on the difference between the actual path of output measures and those generated by a counterfactual cannot be made until long after the event, and estimates based on highly detailed micro damage assumptions will also likely lag behind the event by a considerable margin.

Case in Point: Adding up Sandy

Superstorm Sandy in 2012 provides an interesting case study of how damage estimates from a disaster evolve, starting from the time of the storm. At the outset, EQECAT, the catastrophic risk management firm, issued an estimate on October 29, 2012, prior to landfall, that Sandy would inflict $5 to $10 billion in insured damage and total "economic losses" of $10 to $20 billion (http://www.reuters.com/article/2012/10/29/us-storm-sandy-losses-idUSBRE89S0RG20121029). EQECAT updated its estimate after landfall to $10 to $20 billion in insured losses and economic losses of $30 to $50 billion (http://www.businessinsurance.com/article/20121101/NEWS04/121109990?tags=|306|64|305|76|83|302). These estimates were aggregates for all regions hit by the storm. During the following week Moody's Analytics issued detailed estimates of damage ($30 billion) and output loss (approximately $20 billion) by industry and state.

Throughout November 2012, economists across the affected region engaged in numerous discussions about gauging the impact of the storm; among the groups involved were the New Jersey Treasury Department, the Federal Reserve Banks of New York and Philadelphia (as well as the Board of Governors), the Port Authorities of New York and New Jersey, and some of the universities in the area. These discussions were held while New York and New Jersey officials were preparing estimates of damages that were sent to Congress for consideration in the relief legislation that was ultimately enacted in January 2013. On January 30, 2013—three months after the storm and just after the passage of the relief legislation—the BEA released an estimate that Sandy inflicted $44.4 billion in physical damage. As noted above, the BEA's concept of storm damage is somewhat limited, which helps explain why this number is smaller than some other damage estimates, including the sums reported by New York and New Jersey.[2]

---

2 The BEA's estimate can be found at http://www.bea.gov/national/pdf/sandytable.pdf. The updated damage estimate is $44.9 billion. Annual estimates of losses from events

In December 2012 Rutgers University hosted a conference on regional economic issues, and Sandy's effects were a major topic. In January, economists at Rutgers issued a study of the effects of Sandy (Mantell et al., 2013); they estimated that the storm directly reduced New Jersey's state output by $11.9 billion in the fourth quarter. This loss—not at an annual rate—would be quite a visible share of New Jersey's annual gross product of roughly $500 billion. The report also suggested that there was $4.3 billion in rebuilding activity in the fourth quarter, which would have offset a portion of the direct output loss.

The Rutgers study received a good deal of attention. However, the output loss estimates were apparently based on highly indirect inference. Specifically, the estimates of the direct output loss, prior to any rebuilding offset, were derived under the assumption that storm interruptions reduced weekly output in the state (averaging a bit less than $10 billion) by 2/3 in the first week, 1/3 the following week, and 1/6 the third week. Thus, the state was assumed to lose 1 1/6 week's production, none of which was offset in later weeks after the restoration of power and more normal conditions to the bulk of the state. The assumption about the pace of rebuilding activity was based on the official state estimate that the cost of repairing the physical damage from the storm was $29.3 billion (http://www.state.nj.us/governor/news/news/552012/approved/20121128e.html) (the state also included in its estimates the sums needed for future infrastructure upgrades), and that the rebuilding process would be completed in approximately two years.

While the basis for weekly direct output effects used in the Rutgers study was not specified, the most likely motivation was from observations on the labor front. Most New Jersey residents missed some work during the week of the storm, and there were ongoing problems for several weeks afterward, mainly related to power and transportation, while in some instances flood damage limited access to buildings. Some workplaces in Jersey Shore towns were destroyed, and employers—if they resumed business—were delayed in reopening due to the need to find new sites. There were 154,476 new claims for unemployment insurance filed in New Jersey in November 2012, which was nearly twice as high as the previous top month in the state's history; this number suggested the storm had some significant labor market impact.

Despite these signs of notable labor market stress, only a small fraction of the new claims filed resulted in actual unemployment insurance payments, and the labor market data show that private-sector employment in New Jersey fell by only 4,500, or 0.1 percent, in November 2012. For the fourth quarter as a whole—total and private—employment in the state rose at a higher rate than in either the second or third quarters. Furthermore, the BEA estimated that in the fourth quarter of 2012 wage losses in New Jersey resulting from interruptions caused by Sandy were only 0.7 percent of the state's total—an amount equal to about $400 million (http://www.bea.gov/newsreleases/regional/spi/spi_newsrelease.htm).

such as Sandy are in the BEA fixed asset tables 1.7 and 1.8. GDP table 5.10U contains estimates of quarterly capital transfers stemming from disasters.

It is arguable if the employment and wage numbers mask output effects more on the lines the Rutgers group assumed. Possibly, hours actually worked as opposed to hours paid, slumped dramatically as New Jersey residents stayed home in the dark but remained on their employers' payrolls. In effect, productivity may have dropped, bringing down output. Also, remediation and rebuilding activity could have offset a larger portion of the depressing effects of the storm than the Rutgers study assumed. Increases in employment in construction and administrative and waste management services in late 2012 (two sectors likely to have benefitted from remediation and reconstruction efforts) did more than offset losses in accommodation and foodservices (a sector heavily affected by the storm).

New Jersey's revenues may also provide some insight into the immediate output effect of the storm. Table 9.1 shows that state revenues in November and December 2012 were roughly $160 million below the New Jersey Treasury Department's targets of over $4 billion, or less than four percent short (the storm's negative economic impact was likely more concentrated in November than in December, but numbers of important revenue sources, particularly sales tax, are collected and reported with a one-month lag). If the approximately 4 percent revenue shortfall was entirely due to Sandy reducing nominal state output by that fraction in those months, the loss of output in New Jersey in the fourth quarter would have been around $3 billion, based on monthly state output averaging roughly $40 billion. This is substantively less than the Rutgers estimate of a total output loss greater than $7 billion (the difference between their estimate of the direct disruption effect and the gain from rebuilding).

Using revenue shortfalls to gauge the immediate output loss from Sandy is somewhat problematic. At all times, the linkage between current tax collections and current output is fairly loose. Furthermore, revenue shortfalls in September and October also averaged about four percent, which is comparable to the November–December average. Hence, while revenue results may suggest that state output was less than (implicitly) projected in all four months, they do not obviously imply that the average gap was wider in November-December than in September-October. The failure of the revenue gap to widen after Sandy may be viewed as evidence of a fairly muted immediate output loss from the storm. Thus the combination of little or no employment loss and only modest revenue losses suggests that the Rutgers number could have been an overstatement of the output loss from the storm, even taking into account the possibility that remediation and reconstruction activity stepped up briskly in its aftermath.

Another crosscheck on output losses is provided by a survey from the Federal Reserve Bank of New York which found that only 36 percent of downstate New York manufacturing firms experienced revenue losses due to the storm (http://www.newyorkfed.org/survey/empire/empire2013/2013_06Report.pdf). The average revenue loss in this fraction was 7.2 days, or perhaps 3 percent of a year's total. This suggests an aggregate loss of revenue for manufacturing in the New York state region most affected by the storm of roughly 1 percent of a year's total. Assuming that this ratio can apply to output, and would be applicable to all New

Jersey businesses, these results might suggest an output loss to the state on the order of $5 billion.

Table 9.1 New Jersey state revenues (thousands)

	Revenue		Variance	
	Targeted	Actual	Amount	Percentage
September 2012	$2,363,300	$2,288,177	($75,123)	(3.2)
October 2012	$2,003,000	$1,914,170	($88,830)	(4.4)
November 2012	$1,703,600	$1,519,912	($183,688)	(10.8)
December 2012	$2,342,100	$2,367,682	$25,234	1.1
January 2013	$2,982,200	$3,057,682	$75,482	2.5
February 2013	$1,648,300	$1,804,608	$156,308	9.5

Source: Monthly releases on state revenue collections, New Jersey Department of the Treasury (http://www.state.nj.us/treasury/). The table compares "actual" revenues to "cash budget" figures. The figures are taken from releases issued from October 2012 through March 2013 that report the results for the prior month.

The fairly modest output losses suggested by employment and revenue numbers, and the New York Federal Reserve Bank manufacturing survey, remain a bit puzzling. The reality was that for a large portion of November 2012, much of New Jersey was without power and there were considerable transportation problems stemming from gasoline shortages and closed commuter rail lines.

One potential reconciliation of the scale of disruptions with what appears to be fairly limited losses to aggregate output is provided by the observation that approximately one-third of New Jersey's state output is produced by industries whose quarterly output (and employment) would be relatively insensitive even to a marked interruption by a natural disaster: real estate, education, health care, social welfare, and government. Real estate output is primarily the shelter services of the housing stock, which, except for the homes destroyed or heavily damaged by the storm, was sustained through the period (the discomfort people felt sitting at home after the storm was due to lost power). In education, health care, and government, services not provided during the storm were likely fairly quickly regained: classes and doctor appointments were rescheduled; permits and drivers licenses were issued as offices reopened, back mail was delivered, and so on. In many other instances it is reasonable to assume that much of the lost activity could have been recovered without a particular extension of hours. For example, university faculty may have devoted more time to research during the disruption period, allowing for a pickup in teaching when conditions normalized without extending their working hours.

It is also conceivable that many people were able to do a good deal of work at home during November 2012, even in the absence of power.[3] Fundamentally, utilities are an intermediate input to production. Loss of an intermediate input implies a reduction in multi-factor productivity. A decline in multi-factor productivity does not automatically entail a decline in output; the same amount of output can be produced, with some loss of efficiency, by substituting other inputs for the one whose scarcity has increased. In November 2012 it was possible that many New Jersey workers were able to do much of their work from home using their own technology. In effect, labor (home time) and capital (home technology) may have been substituted for unavailable inputs (office technology and power, and transportation). Residential utility outages were not necessarily an insurmountable obstacle to many New Jersey residents getting some work done. Virtually all New Jersey residents had access to places such as public libraries where power was available and, of course, many residents made use of home generators (though obtaining gasoline for the generators was very difficult in large parts of the state).

Since the wave of reports in the wake of the storm there have been few updates relevant to estimates of Sandy's aggregate cost, either to the nation as a whole or to New Jersey in particular. On the output front, a US Commerce Department study estimated that the storm reduced spending on travel and tourism in New Jersey by $828.6 million in the third quarter of 2013, leading to an output reduction of $1.2 billion (US Department of Commerce, 2013). While a substantial loss to the sector, given the size of the state's economy (over $500 billion in annual output) this is fairly modest, and much less dramatic than the Rutgers report may have suggested. Another report on travel and tourism in the state, using data on physical activity measures in the summer of 2013, suggests that even these estimates were high (New Jersey Economic Development Authority, 2013). In any event, neither report substantively addressed the issue of a more immediate output loss at the time of the storm.

Improving Aggregate Disaster Estimates

The major problems in estimating aggregate disaster costs by region are receiving timely, definitive figures on the value of property damage by geographic area, and estimating the output effects, especially the size and extent of any initial disruptions. The sharp divergence between initial estimates of a very large negative

3 Remarkably, US Department of Energy figures show that November 2012 was the only month in the period from October 2012 through January 2013 in which New Jersey electricity usage was higher than a year earlier. This is scarcely believable, and may be an artifact of utilities' inability to gather meter readings in a timely fashion as a result of Sandy. Still, the scale of electricity usage declines in the state in the whole period around the storm does not appear to be consistent with an output loss of the magnitude suggested in the Rutgers report.

effect on the output of New Jersey and what now appears to be very limited effects on the state's overall activity suggests that extrapolating limited labour market information to infer output losses is problematic. Apparently, output was more robust than was suggested by the immediate scale of disruption.

Unfortunately, data are not available to explore hypotheses about the nature of production workarounds during disasters. To gain more understanding on this issue, it may be useful to undertake time studies during such periods: What, exactly, did New Jersey workers do when their workplaces were closed or inaccessible? If it is inconvenient to do such studies during the aftermath of a disaster, a follow-up as soon as possible when conditions normalize may be desirable.

Such information would have several uses. First, of course, it could help government, business, and household preparations for a disaster: precisely how much work can a person do at such times, and what are the obstacles to working? What sort of home equipment proves useful? Second, getting an idea of how much immediate output disruption occurred as a result of a disaster would help local officials get a grasp on revenue impact. A practical issue is determining what body would undertake such studies. Conceivably, the American Time Use Survey panel may be exploited, but an infrastructure and question set would need to be in place (and funded) to make off-cycle queries.

While there are relatively fewer problems in the area of compiling aggregate physical damage estimates, receiving more timely estimates would be useful, especially in framing requests for aid in rebuilding. Regional officials were well aware that the optimal time for requesting national assistance was in Sandy's immediate aftermath. Such requests necessitate some idea of the monetary volume needed. The BEA providing its estimates on a cycle determined by the timing of the event, rather than on its normal GDP release schedule, may help in streamlining the process of aid requests, especially if damage estimates were provided state-level as well as nationally. Also, as noted, expanding the menu of damage estimates distributed by the BEA or other federal agencies to include losses of personal property and business inventories would further aid the process of developing area needs.

Another enhancement to grasping the scale of the physical cost of a disaster would be more complete data on the proper comparative metrics. Obviously, the sheer current dollar cost of physical damage caused by a disaster is a somewhat inadequate gauge of its significance. What may be more salient (aside from the usual issues about examination of nominal, rather than constant-dollar, costs) is the relative burden thrown upon an area. At the current time the most comprehensive measure of a state's economy is GDP. An example illustrates the effect of such scaling: while the total dollar value of physical damage from Sandy was apparently not radically out of line with that arising from Katrina, a comparison of volumes of aggregate activity in New York-New Jersey with Louisiana-Mississippi indicates that the relative regional impact of the events were very dissimilar. The BEA's estimate of physical damage caused by Katrina is $110.4 billion, or about 2.5 times the estimate for the damage caused by Sandy. However, the combined nominal GDP of New Jersey and New York was $1.64 trillion in 2011, while the combined GDP of Louisiana and Mississippi was $278.3 billion in 2005. In other words, Katrina's damage equaled about 40 percent of a year's output for Louisiana and Mississippi, while Sandy's damage amounted to less than three percent of the output of New Jersey and New York.

Comparing damage to the stock of physical assets with the flow of GDP is limited, since differing levels of capital intensity have a bearing on the magnitude of the loss. If a region has a relatively small capital stock, a loss of capital equaling the same fraction of GDP as in an area with a larger stock of capital may be more damaging. Thus, a more meaningful metric of the scale of a disaster might be the ratio of physical damage to the value of reproducible assets in the affected region. Unfortunately, there are no series that report regional asset values on a consistent basis. However, it seems that from other existing data reasonable approximations can be produced. For example, Table 9.2 uses national data on the value of reproducible assets, combined with data on New Jersey economic activity, to produce an estimate of the value of reproducible capital in New Jersey at the end of 2011. It is assumed, consistent with the methodology of Garofalo and Yamarik (2002) and Yamarik (2013) that the value of reproducible capital used in a New Jersey industry is the same fraction of the national total, as is the industry's output. Unlike these other studies, the capital stock total includes housing (the real estate industry can be considered as the owner of all housing). The value of durable goods owned by New Jersey residents is assumed to be the same fraction of the national total as is New Jersey residents' share of the national total of net earnings by place of residence plus transfer payments to persons.[4]

With these assumptions, the value of reproducible capital in New Jersey at the end of 2011 was estimated to be $1.7 trillion (the corresponding national total was $51 trillion; in other words New Jersey appeared to have had somewhat over three percent of the nation's stock of reproducible capital, which is, not surprisingly, comparable to the state's share of national output). Sandy's damage to physical assets in New Jersey (including the cost of reconstruction) is estimated to have been in the neighborhood of $30 billion, or equal to over 1.5 percent of the value of capital in the state. While the estimate of New Jersey's capital stock is, of course, very approximate, it is clear that the value is neither under $1 trillion nor higher than $3 trillion. This rough estimate probably gives a better idea of the scope of the damage than do mere dollars, or even a comparison to state GDP.[5] The sudden loss

4 Overall personal income, an obvious candidate for such a scaling, includes imputed rent from owner-occupied housing. Since owner-occupied housing estimates are implicit in the industry counts, it is logical to use a less comprehensive series for the scaling of durable goods. Net earnings by place of residence comprise the major share of personal income and includes out-of-state earnings, which is a significant item for New Jersey residents.

5 These estimates do not take into account the recent categorization in the NIPA of R&D spending as final demand, which will also swell capital stock estimates at the national and state levels. In many ways this change will be useful in judging the scale of losses in

of—or sudden need to rebuild—1.5 percent of the capital stock of a state is a major event, and for a state as large as New Jersey it is noticeable on the national scale, even before adding in the damage to New York and other states (which appear to have been in total somewhat larger than New Jersey's).

Table 9.2 New Jersey reproducible capital, year-end 2011 (billions)

Business and Government Total	$1,604.3
Real Estate	$817.5
All Other Private	$481.7
Government	$325.0
Consumer Durables	$170.4
Total Reproducible Capital	$1,774.7

Source: Author's computations, based on US Bureau of Economic Analysis National Income and Product Account data on fixed capital, national and New Jersey industry outputs, and national and New Jersey personal income components. Data used are prior to the July 2013 revision of the National Income and Product Accounts.

Conclusion

It is important to clarify what is meant by the "cost" of a disaster. For welfare purposes, physical destruction of reproducible assets is important and likely the main component, as least for events which do not entail widespread loss of life. From the point of view of near-term income and revenues in the affected area, as well as the potential impact on national economic numbers, output interruption and loss may be the more useful metric. Adding the two together is useful for some purposes—for instance, private insurance often covers business interruption as well as physical damage, and other forms of disaster relief will ameliorate both problems—but they are conceptually different.

We have little understanding of the aggregate size of output interruptions of typical natural disasters, and, most importantly, how they may correlate with rough measures of change in market labor inputs at such times. It is possible that some effort to track actual work effort through mechanisms such as time surveys would help in getting a better idea of actual worker experience in such periods.

Timelier estimates of damage would certainly assist in understanding the dimensions of the event and framing requests for assistance, and taking some steps to enhance regional data systems would also aid in understanding the relative size of events.

---

disasters. In the last analysis, R&D capital is knowledge. Knowledge is essentially not lost in disasters, unless there is wide-scale loss of life.

Appendix: Sandy's Potential Impact on Inventory Investment

Media and other reports have suggested that New Jersey businesses and households lost as many as 30,000 automobiles as a result of Sandy. This total includes imported autos destroyed on the New York harbor docks, which are located in New Jersey. Apparently as much as 80 percent of the autos lost were the property of households. If the average value of what appears to be the roughly 6,000 business autos lost-new cars on dealer lots, and those on the docks was $30,000, the loss would have been about $180 million, which would have been treated in Q4 GDP as a reduction in inventory investment at an annual rate of $720 million. Other business inventories would have also been lost to the storm. However, even taking these into account it is a little hard to think that storm damage accounted for more than a small portion of the very large drop in the annual rate of inventory accumulation between the 3rd and 4th quarters of 2012 reported in the National Income and Product Accounts (the estimate as of early 2014 was $68.6 billion). The closure of the Port of New York for a period after the storm, and the difficulties in transportation in the region, could well have contributed to the inventory investment decline by reducing the inflow of imports, but such would have no net effect on GDP.

New Jersey also contains a significant portion of the nation's logistics infrastructure. Contacts in the logistics sector report that power outages created severe problems at many New Jersey warehouses and distribution centers following Sandy. In such circumstances it is arguable that these firms' reports on inventory levels were less reliable than normal, which could have affected the national Q4 inventory investment estimates, but it's not clear in which direction.

Acknowledgments

The author would like to thank the many people with whom he discussed the impact of Sandy, including Jason Bram, Alexander Heil, Michael Lahr, Len Nakamura, Mike Trebing, Jim Orr, and Joe Seneca, as well as seminar participants at the Federal Reserve Bank of New York. The views expressed in this paper are those of the author, and need not reflect those of the Department of the Treasury or the government of New Jersey.

References

Baritto, F. (2008). *Disasters, Vulnerability and Resilience from a Macro-economic Perspective: Lessons from the Empirical Evidence.* New York: United Nations International Strategy for Disaster Reduction (November).

Blomberg, S.B. and D.H. Gregory (2009). Estimating the Macroeconomic Consequence of 9/11. *Peace Economics, Peace Science and Public Policy* 15(2): 1–26.

Bram, J., J. Orr, and C. Rapaport (2002). Measuring the effects of the September 11 attacks on New York City. *Federal Reserve Bank of New York Economic Policy Review* 8(2): 5–19.

Cochrane, H. (1997). Economic impacts of a Midwestern earthquake. *NCEER Bulletin* 11(1): 1–5.

Garofalo, G.A. and S. Yamarik (2002). Regional convergence: Evidence from a new state-by-state capital stock series. *The Review of Economics and Statistics* 84: 316–323.

Greenberg, M., M. Lahr, and N. Mantell (2007). Understanding the economic costs and benefits of catastrophes and their aftermath: A review and suggestions for the U.S. Federal Government. *Risk Analysis* 27: 83–96.

Greenberg, M., P. Lioy, B. Ozbas, et al. (2013). Passenger rail security, planning, and resilience: Application of network, plume, and economic simulation models as decision support tools. *Risk Analysis* 33: 1969-1986.

Guimaraes, P., F.L. Hefner, and D.P. Woodward (1993). Wealth and income effects of natural disasters: An econometric analysis of Hurricane Hugo. *The Review of Regional Studies* 23(2): 97–114.

Mantell, N.H., J.J. Seneca, M.L. Lahr, and W. Irving (2013). *The Economic and Fiscal Impacts of Hurricane Sandy in New Jersey*. New Brunswick, NJ: Rutgers Regional Report Issue Paper Number 34, January.

New Jersey Economic Development Authority (2013). *New Jersey: Stronger Than the Storm. Recap Report, December 2013*. Trenton, NJ: NJEDA. http://www.njeda.com/web/pdf/STTSReport.pdf.

Pelling, M., A. Özerdem, and S. Barakat (2002). The Macro-Economic Impact of Disasters. *Progress in Development Studies* 2: 283–305.

US Department of Commerce, Economics and Statistics Administration, Office of the Chief Economist (2013). *Economic Impact of Hurricane Sandy: Potential Economic Activity Lost and Gained in New Jersey and New York*. Washington, DC: US Department of Commerce. http://www.esa.doc.gov/sites/default/files/reports/documents/sandyfinal101713.pdf.

Xiao, Y. and S. Van Zandt (2012). Building community resiliency: Spatial links between household and business post-disaster return. *Urban Studies* 49: 2523–2542.

Yamarik, S. (2013). State-level capital and investment: Updates and implications. *Contemporary Economic Policy* 31: 62–67.

Chapter 10
Conclusion: Stages and Scales of Urban Resilience Strategies

Gary Sands, Pierre Filion, and Mark Skidmore

The book provides a compendium of chapters, each of which is motivated by recent natural disasters affecting urban areas, with the exception of one historical catastrophe. The cases cast light on responses directed at four distinct stages of natural disasters and their consequences: preparedness; interventions taking place during and immediately after the disaster event, temporary shelter and other relief efforts; and long-term reconstruction (see Harrald, 2006). In their interpretation of each of the respective cases, the different chapters emphasize the role of specific stages.

The disaster events described in this volume represent a wide range of circumstances and contexts, ranging from seasonal flooding in a developing country to major storms and earthquakes in highly industrialized nations. Even when the precipitating events are similar, the consequences can be disparate; hurricanes Katrina and Sandy had very different impacts in part because of socio-economic differences between New Orleans and New York, as well as because of differences in preparedness. While Japan has suffered from numerous earthquakes and a tsunami, the disaster that led to the Fukushima nuclear plant meltdown was singular because of the long-run devastating implications of the ongoing radioactive leak. There is no one size fits all disaster.

And there is certainly no one size fits all disaster recovery scenario. As a result, it is simply not possible to conceive of a disaster rubric that adequately anticipates all possibilities. The unique characteristics of each disaster event, as well as the uncertainty of the timing of occurrences, makes it impossible to develop a 'check list' that can be applied to any and all disasters. Nevertheless, based on the four stages of reaction to a disaster, it is possible to establish a set of guidelines that provide positive results in a variety of disaster events.

1. Preparation

Communities and governments must begin planning their responses to disasters well before such events occur. The extent and content of these preparations will depend to some degree on the nature of both the community and the type of disaster anticipated. Some disaster events (earthquakes, for example) typically occur suddenly, while others, such as storms, may provide some advance warning. The size and the resources of the community will dictate the type of preparation that is feasible.

One element of preparation addresses the emergency responses in the immediate aftermath of the disaster event. Emergency supplies and equipment, ranging from food and drinking water to transportation, medical services and heavy equipment to clear debris, must be identified. Even if these items cannot be stockpiled ahead of time, their location, and the means to deliver them where needed, must be identified. Similarly, the necessary personnel for dealing with the disaster must be identified, trained in advance and prepared to quickly assemble when and where they are needed.

An important part of the preparation phase focuses on mitigation, taking the steps necessary to limit the extent of damage when a disaster does occur. Seawalls and levees are obvious examples. Land use planning can also be used to minimize damage by preventing development in high-risk areas (Masterson et al., 2014). Development regulations that include seismic codes and fireproof materials should be put into place where appropriate (Atkinson, 2014).

In many cases, the technological solutions that can greatly reduce the damage caused by natural disasters already exist; however, much of today's built environment predates these advances. Going forward, a community can insist that new development complies with the most appropriate standards. But this does not address the problems of existing structures. The cost of retrofitting every public and private structure that does not meet modern disaster resistance standards is beyond the means of even the richest community (see, for example, Atkinson, 2014). Thus, an important component of preparation is determining where, when and how much retrofitting should be undertaken. In many instances economic and political circumstances limit the application of these technological advances. For example, in poor countries it is difficult to impose strict building codes because of the limited economic means of the population and a weak regulatory framework.

A number of chapters expose the failure of preparedness and its consequences. Most cases chronicled in the book performed poorly in terms of preparedness. But two examples of relative success in assuring preparedness stand out among the cases discussed in the book: the Ghana community of Salon Zongo and Hurricane Sandy. It is noteworthy that both examples benefitted from circumstances that were particularly favorable to preparedness. In the first instance, tight social ties and strong leadership were conducive to community mobilization around preparedness. There was also the predictable nature of the hazards for which the community maintained preparedness—repeated flooding. Regarding Hurricane Sandy, the Obama administration was strongly motivated to avoid a repetition of the damaging political sequels that resulted from the deficient Hurricane Katrina rescue assistance (Johnson and Olshansky, 2013). The imminent presidential election made all the more pressing the need to demonstrate competence in matters of disaster relief.

But planning and preparation are not the only ways to achieve a successful disaster recovery. Two of the more successful recovery cases are noteworthy for their limited pre-event planning. In the Lisbon earthquake of 1755 the relative lack of pre-event preparations seems to have been at least partially offset by the decisive response from a national government that provided significant resources for recovery efforts led by the Marques de Pombal. The location of the disaster, the capital of a relatively small, homogenous and prosperous country, also helped. The success of the Lisbon recovery efforts, still a model for disaster recovery 270 years later, stands in sharp contrast to the efforts to rebuild London after the Great Fire of 1666 (Tinniswood, 2003).

The other noteworthy case is the Christchurch earthquake, chronologically one of the most recent events described in this volume. Like Portugal in the 18th Century, New Zealand is a relatively small, prosperous nation. While the amount of pre-disaster planning and preparation was typical, the relative success of the recovery efforts was a result of administrative decisions taken after the earthquake. Recovery efforts were facilitated by the establishment of coordinating bodies that included a wide range of stakeholders, including national and local government agencies, construction and insurance companies. The result was a recovery implementation process that was relatively efficient and expeditious.[1] While higher levels of planning might have improved the process still more, the Christchurch experience provides another model of how post-disaster actions can be effective.

The content of the book in this matter mirrors the interest expressed in the literature for preparedness. Among other things, the literature presents surveys of preparedness for natural disasters (Tierney, Lindell and Perry, 2001), while concentrating on hazard mitigation measures (Godschalk, 2003), and the need for social learning around issues related to natural disasters (Norris et al., 2008; Paton and Johnson, 2001). Research also centers on the planning approaches that are best suited to preparedness (Auf der Heide, 2006). In this latter respect, Burby (1998) discusses the land use planning models that can minimize the impact of natural hazards.

2. Response

The immediate response in the aftermath of a disaster event represents a second crucial phase. Here the emphasis shifts from activities that must be well planned in advance to those that must be well executed. If the necessary food, water, shelter and medical care cannot be delivered in a timely and effective manner, previous planning efforts will be of no avail. The restoration of utilities and public services, even on a temporary basis, may be an important part of the response phase as well.

1 New administrative structures were created in Louisiana after Hurricane Katrina, but with much more limited effect. See Chapter 7.

The case studies described in this volume present a variety of response scenarios, most of which were less than adequate. Probably the two best examples of dealing with disasters are the earliest (Lisbon, Portugal) and most recent (Christchurch, New Zealand). There are some indications of institutional learning; for example, the Katrina driven changes in policies that facilitated more effective reactions after Hurricane Sandy. Yet the Kobe, Japan earthquake does not appear to have informed activities after the Bam, Iran earthquake.

Two events related in the book point to the consequences of ill-fated decisions taken during or immediately after a disaster. Some chapters refer to Hurricane Katrina as an example of inadequate preparedness and, especially, an ill-conceived and poorly deployed evacuation strategy. Better preparedness and decision making during and after the hurricane could have significantly reduced the casualties and suffering it caused. The other event is the Fukushima nuclear meltdown, following the earthquake and ensuing tsunami. Again, the combination of wanting preparedness and ill-informed decisions regarding the handling of a nuclear plant in distress resulted in severe damage to the plant and a considerable release of radioactivity in the environment. The lesson emerging from these examples is that, because decisions must be taken promptly in the heat of the action, they require clear channels of command, which are well supplied with the information required to make these decisions. Also, in these circumstances there is little time to consider and debate the consequences of different decisions; such deliberations should be incorporated into the preparedness process.

3. Early Recovery

The third phase of the process deals with efforts leading to the resumption of "normal," pre-disaster levels of activity. While there are no hard and fast guidelines distinguishing Response from Early Recovery, the difference can be likened to the distinction between 'search and rescue' efforts and 'search and recovery' efforts. Emphasis shifts from saving lives to providing for a resumption of basic social and economic functions.

Activity in the Early Recovery period is shaped by the characteristics of the damaged area. In both Lisbon and Christchurch, the disaster events struck the respective city centers. The Early Recovery period was extended well past the debris clearance stage to allow for adequate planning of the long-term reconstruction. This required the expenditure of considerable resources before the actual reconstruction could begin. In contrast, the more limited resources available in Bam resulted in short-term developments becoming permanent.

Planning for this phase should begin during the Response phase, taking into consideration the relevant information about the disaster event. Some time may be required for information gathering; for example, determining which buildings can be repaired and which require demolition.

The cases of Kobe in Japan and Bam City in Iran illustrate the need for a quick availability of temporary shelter and other necessities of life such as food, water, health care and schooling. These services must be provided immediately after the disaster (Hosseini et al., 2009). The psychological needs of victims, often in a state of trauma after the loss of family and friends, must also be cared for. The two cases raise the difficulties of rapidly creating shelter camps that provide an acceptable quality of life. In both cases, accessibility to stores and services, jobs, family and friends proved to be deficient. The discussion of temporary shelters also points to the danger that, when reconstruction efforts are too slow or even absent, temporary shelters will become permanent forcing their residents to depend only on very basic necessities of life. There is also the need to balance the distribution of resources available for recovery between temporary shelter and permanent reconstruction (Bolin and Stanford, 1991).

4. Long-Term Reconstruction

The last phase of the disaster process requires careful planning to determine the nature of the redevelopment that will occur as well as how the reconstructed area will relate to the surrounding areas. Incorporating mitigation strategies into reconstruction activities is an important consideration. In some instances, this might include the abandonment of the affected area and the prohibition of any reconstruction. For example, a Japanese coastal village was relocated three kilometers inland after the 2011 disaster (*The Economist*, 2014).

The literature suggests that urban recovery is common, notwithstanding cities that were already in a state of decline before the disaster struck (Vale and Campanella, 2005; Vigdor, 2008). The case of New Orleans described in Chapter 7 highlights problems in bringing evacuated residents back to the poorer parts of the city, such as the Ninth Ward, which had been losing population long before Hurricane Katrina. While urban recovery is the norm, it can take different forms. The main dilemma confronting decision makers is whether damaged areas should be returned to their pre-disaster condition, or rebuilt according to more resilient planning models and building codes (Burby et al., 2000). The problem with the first option is that the city would remain vulnerable to a repetition of the type of event that caused the damages from which it is attempting to recover. The second option addresses this problem but is itself confronted by public attachment to the pre-disaster cityscape, limited resources, and desire to rebuild as quickly as possible. The cases of Lisbon and Christchurch, introduced in chapters 5 and 6, demonstrate how these two cities have launched reconstruction programs, 250 years apart, aiming at enhancing urban resilience in the face of earthquakes. The two chapters also point to the obstacles encountered by these reconstruction strategies.

Political and economic considerations cut across all four phases of disaster activity. In advanced economies, like Japan and New Zealand, commercial insurance can supplement government disaster relief spending. The National

Flood Insurance Program in the United States may create a situation of "moral hazard," which can increase risks and add to disaster losses.

The examples of disasters in poorer countries, such as Ghana and Iran, highlight the difference between knowing what to do and actually being able to do it. Rebuilding to current seismic standards (Bam) or implementing flood control measures (Salon Zongo) are unlikely given local resources, both public and private. This is quite a different situation than one in which neoliberal principles limit the role of government.

These neoliberal principles can be overcome by political considerations. The political necessity to be seen doing something rather than standing by idly, has been a motivating factor of the governments in power in Japan and the United States. In both instances, the perception of inadequacy in responses to Fukushima and Katrina has shaped subsequent government actions. But simply increasing budgets does not necessarily result in optimal outcomes. The US governments response to Hurricane Sandy was much more extensive and expensive than its response to Hurricane Katrina, but not all of the deficiencies were addressed (Johnson and Olshansky, 2013; Waugh, 2007). Shortly after the 2011 disaster and its disastrous response, the Japanese government pledged to spend almost $10 billion on constructing seawalls and breakers (*The Economist*, 2014). Over time, questions have been raised about the adequacy and appropriateness of this plan; nevertheless, it demonstrates the government's commitment to action.

Together, the chapters present a stage-wise perspective on natural disaster relief and recovery. Each of these stages are shown to be important to the outcome of disaster responses, and each demands its own specific approach. For example, the resources and skills required to assure preparedness are different from those needed during the long-term reconstruction phase. And the very quick reactions that should take place during and immediately after the disaster must rely on a decision-making process that contrasts sharply with the long-term planning encountered at the preparedness and reconstruction stages. The cases have demonstrated the importance of all of these stages. Moreover, the effects are cumulative: insufficient preparedness has caused later stages to falter. Misinformed decisions taken during the disaster have exacerbated its consequences and caused unnecessary suffering and damage. The quick setting up of temporary shelters and other infrastructures and services provides the necessities of life essential to the survival of victims and their eventual return to restored areas. Finally, careful reconstruction planning can improve resilience and therefore reduce vulnerability to future natural disasters.

Not only can the content of the book be divided in different stages, but it can also be fitted into two scales at which interventions are implemented: the scale of the state, its levels and agencies, and that of community organizations. The chapters are replete with examples of actions, at all stages of the recovery process, originating from the state. For example, the case of the rebuilding of Lisbon is indeed introduced as a prototype of state reconstruction strategies, and the state is present in nearly all other cases discussed in the book. These chapters thus *de facto* present governments as essential actors at all stages of responses to natural disasters. But the cases also indicate that the participation of the state does not necessarily translate into successful preparedness and recovery (see also Stehr, 2006). Many of the more recent disasters chronicled in the book were not the object of well-coordinated, forceful and effective government actions such as those that benefited Lisbon after the 1755 earthquake. Frequently, state responses lacked preparedness, resources and coherence. As noted in the introduction to the book, these circumstances can be linked to the expected tendency to give priority to immediate demands related to high-probability events rather than to eventual hazards, where probabilities are difficult to gauge. This largely explains the role insufficient preparedness plays in difficulties encountered by relief and recovery efforts. The introduction described how this tendency is also consistent with the tension inherent in the distribution of resources between the public and private sector, and the priority neoliberalism gives to the market over state-driven planning.

The other scale pertains to local, community-based efforts. These are the main, if not the only, source of preparedness, relief and recovery when the presence of the state is absent or minimal. We encountered such a situation in responses to repeated floods in the Ghanaian community of Salon Zongo described in Chapter 2. Community relief can also fill gaps in more elaborate state interventions. Such relief can take place before state agencies have time to mount responses, and address needs, such as caring for a neighbor in distress, that can easily be overlooked by large-scale efforts. Chapter 4 notes the part residents played in bringing assistance to a large number of Kobe earthquake victims before the arrival of a more organized rescue. The advantages of community efforts relative to elaborate top-down strategies reside in an ability to react quickly and fine-grain knowledge of local conditions (Comfort, 1999; Comfort and Haase, 2006).

There are also limitations to community-based efforts, however. Community responses cannot rely on sophisticated rescue equipment and specialized knowledge, especially in medical matters. After a disaster, community self-help can provide immediate assistance in the form of food, water and first aid to close-by people in need. But for many other needs, such as the supply of the essentials of life to a large population in need, the rescuing of people caught deep in the debris of destroyed buildings, and the quick provision of emergency shelter, the involvement of better resourced, equipped and organized agencies is required. Furthermore, community self-help cannot be improvised. As Chapter 2 demonstrates, it must rest on pre-existing social networks and requires leadership that sustains public awareness of the risks of disasters and the need for readiness (see also Buckle, 2006). Community self-help cannot be switched on and off according to need.

Natural hazards can never be completely prevented and will remain largely unpredictable with respect to their timing and severity. And their ferocity will occasionally overwhelm the most robust human defenses. This is one reason such disasters will never be fully overcome and will remain a threat to cities. Even when the existence of disaster risk is known, there is no assurance that the potential damage that may result is fully appreciated. There may also be a tendency to expect that previously adopted measures, such as building codes and flood control structures, will be adequate in the face of future disasters; this may provide a false sense of security.

Planning is an essential element with respect to each stage of the process—preparation, response, early recovery and long term reconstruction. Although relevant planning methodologies have been developed (for example, Masterson et al., 2014), actual planning responses are likely to prove inadequate because the planning efforts that are required involve forward decision making under conditions of extreme uncertainty. Inevitably, even the best plans will result in mistakes and missed opportunities. As communities work through a disaster event, options will be foreclosed and new opportunities will arise. Adaptation and flexibility will inevitably be required.

As the chapters of this book have illustrated, a great deal can be learned from past disaster events. At the same time, there are limits to what can be learned from the historical record. Not only is each disaster event unique, but the contexts in which these events occur are also different. Social, economic and political circumstances affect the range of appropriate responses. While many of the lessons learned from the Lisbon earthquake remain valid today, the response and recovery to the Angra do Heroísmo earthquake was in many respects less successful, not because of a lack of knowledge but because of the remoteness, relatively low income and lack of institutional resources in the Azores.

The notion of effective policy transfer between industrialized and developing nations is also problematic. The US Federal government obviously learned important lessons from Hurricane Katrina that were successfully applied to Hurricane Sandy. But the differences between the Bam City and Kobe earthquakes highlight the relevance of context. Shortcomings in response and recovery efforts can occur because of a lack of knowledge, poor judgments or simply a lack of sufficient resources to implement known solutions.

Another factor that we must consider with regard to the enduring natural hazard risks to cities is the nature of societal choices about the distribution of resources. There will always be a limitation on the resources that can be directed at the prediction and mitigation of natural disasters, and at recovery efforts. Dealing with possible or actual disasters requires a reallocation of funds from other purposes. Hence, there are perennial limits on the resources available to confront disasters. The way these limitations are determined and their severity mirror prevailing political and economic regimes. The present neoliberalism turn, which gives priority to market economics and mistrusts government-induced planning, makes it difficult to direct more resources towards natural hazard preparedness (Burgess, Carmona and Kolstee, 1997). As a result, it is difficult to determine in advance whether investments in preparations are adequate or not.

Still, there is room within these constraints for learning from the responses directed at past disasters and thus improve future preparedness, recovery and reconstruction. With a goal of extending knowledge on urban recovery from natural disasters, the book has introduced cases pointing to the improvement of natural hazard impact mitigation, recovery and reconstruction. The time is ripe for such a conversation. As population grows and becomes more urban, cities are increasingly reliant on sophisticated technological systems, which are vulnerable to natural hazards, and the climate researchers indicate that the emission of greenhouse gases makes climate events increasingly unpredictable and perilous. There is much to gain in terms of preparedness and recovery from sharing knowledge about cities at risk.

References

Atkinson, W. (2014). Quake Control: A Status Report on Seismic Codes. *Planning* 79(6): 36–38.

Auf der Heide, J. (2008). The economic aftermath of Hurricane Katrina. *The Journal of Economic Perspectives* 22(4): 135–154.

Bolin, R. and L. Stanford (1991). Shelter, housing and resources: A comparison of U.S. disasters. *Disasters* 15: 24–34.

Burby, R. (ed.) (1998). *Cooperating with Nature: Confronting Natural Hazards with Land-use Planning for Sustainable Communities*. Washington, DC: Joseph Henry Press.

Burby, R., R. Deyle, D. Godschalk and R. Olshanski (2000). Creating hazard resilient communities through land-use planning. *Natural Hazard Review* 1: 99–106.

Burgess, R., M. Carmona, and T. Kolstee (1997). Cities, the state and the market: Contemporary macroeconomic strategies and urban policies in developing countries, a critical review, in R. Burgess, M. Carmona, and T. Kolstee (eds) *The Challenge of Sustainable Cities*. London: Zed Books.

Comfort, L. (1999). *Shared Risk: Complex Systems in Seismic Response*. Oxford: Pergamon.

Comfort, L. and T. Haase (2006). Communication, coherence, and collective action: The impact of Hurricane Katrina on communications infrastructures. *Public Works Management and Policy* 10: 328–343.

Economist, The (2014). "The Great Wall of Japan." June 14: 37.

Godschalk, D. (2003). Urban hazard mitigation: Creating resilient cities. *Natural Hazard Review* 4(3): 136–143.

Harrald, J. (2006). Agility and discipline: Critical success factors for disaster response. *The Annals of the American Academy of Political and Social Science* 604: 256–272.

Hosseini, K., M. Jafari, M. Hosseini, et al. (2009). Development of urban planning guidelines for improving emergency response capacities in seismic areas of Iran. *Disasters* 33: 645–664.

Johnson, L.A. and R.B. Olshansky (2013). The road to recovery: Governing post disaster reconstruction. *Land Lines*, July 14–21.

Lynch, E. (2015). Hurricane Sandy's Wake-Up Call. *Planning* 81(2): 26–30.

Masterson, J.H., W.G. Peacock, S.S. Van Zandt et al. (2014). *Planning for Community Resilience: A Handbook for Reducing Vulnerability to Disasters*. Washington, DC: Island Press.

Norris, F.H., S.P. Stevens, B. Pfefferbaum, et al. (2008). Community resilience as a metaphor, theory, set of capacities, and strategy for disaster resilience. *American Journal of Community Psychology* 41: 127–150.

Paton, D. and D. Johnston (2001). Disasters and communities: Vulnerability, resilience and preparedness. *Disaster Prevention and Management: An International Journal* 10: 270–277.

Stehr, S.D. (2006). The Political Economy of Urban Disaster Assistance. *Urban Affairs Review* 41: 492–500.

Tierney, K., M. Lindell, and R. Perry (2001). *Facing the Unexpected: Disaster Preparedness and Response in the United States*. Washington, DC: Joseph Henry Press.

Tinniswood, A. 2003. *By Permission of Heaven*. New York: Riverhead Bppks.

Vale, L.J. and T.J. Campanella (eds) (2005). *The Resilient City: How Modern Cities Recover from Disaster*. Oxford: Oxford University Press.

Vigdor, J. (2008). The economic aftermath of Hurricane Katrina. *The Journal of Perspectives* 22 (4): 135–154.

Waugh, W.L. (2006). The political costs of failure in the Katrina and Rita disasters. *The Annals of the American Academy of Political and Social Science* 604: 10–25.

Index

aftershocks 34, 83, 97, 99, 105
Angra do Heroismo *see* earthquake, Angra do Heroismo
Arga-e-Bam *see* Bam Citadel

Bam Citadel 67–8
Bam City 9, 65, 70, 72–3, 74 *see also* earthquake, Bam City
building codes 33, 65, 164, 173, 206, 209, 212
building materials 194
Bureau of Economic Analysis (BEA) 103, 193, 196, 200
business activity *see* commerce
business inventories 194

Christchurch 9, 97, 99–100, 102–5, 116–18, 122, 126, 157, 129–30, 207–8, 209
  CBD 99, 103, 105, 109, 128
  insurance 118–21
  lifeline infrastructure 111–15, 130, 131
  property market 106–8
climate change 1, 4, 7, 8, 17–18, 44, 53, 155, 163
commerce 85–7, 125
Community Development Block Grant (CDBG) 148–9
community organizations 65, 69, 77–8, 210

density 35, 40, 73, 140, 150, 153–4, 163
developing nations 16, 206, 212

earthquake 1, 2, 3–5, 6, 8, 19, 35–6, 53, 81, 121,163, 165, 170, 189, 205, 209
  Angra do Heroismo 9, 81, 86–8, 89, 91, 206, 211, 212
  Bam City 9, 61, 65, 70–77, 206, 208, 211
  Canterbury 9, 97–105, 107–9, 110–16, 127, 129–31, 206
  economic impact 122–5
  Haiti 61
  infrastructure 110–15
  insurance 118–21
    upgrading 117–18
  Japan 31–2, 33–5, 37, 40–42, 205
  Kobe 9, 35, 38, 53, 61, 64, 65–9, 208, 211, 212
  Kocaeli 61, 63
  L'Aquila 110
  Lisbon 2, 9, 81, 82–5, 89, 91, 206, 211, 212
  Loma Prieta 61
  Northridge 35
  Wenchuan 61
economic loss 118, 150, 195
emergency shelter 9, 66, 211
evacuation 3, 31, 34–6, 53, 66, 147, 208 *see also* relocation

Federal Emergency Management Agency (FEMA) 47–50, 142–3, 147, 170, 193, 195
fire 1, 23–4, 26–7, 34, 35, 56, 65, 70, 82–3, 166, 209
firebreak 85
firefighters 39, 64–5, 70
fireproof 206
firewall 65
flood 1–3, 7–8, 19, 21–7, 31–3, 35–6, 41, 44, 50–51, 103–5, 106, 108–9, 144–5, 163–4, 165–6, 168–70, 174, 205–6, 210, 211, 212
  flood damage 3, 25, 32, 196
  flood mitigation 104, 109, 164, 169
  flood plain 1, 48, 50, 103–4, 164, 168, 174
  flood prone 22, 75, 165

flood wall 145, 169 *see also* sea wall
focus group 22–7
Fukushima Nuclear Plant 8, 31–5, 36–40, 205, 208, 210
lessons 53–4
response 41–2

Great East Japan Earthquake–Tsunami *see* tsunami, Great East Japan
Gross Domestic Product (GDP) 100, 118, 194, 196, 200–201, 203

Hanshim earthquake *see* earthquake, Kobe
heritage buildings 81, 86, 88, 90
hurricane 2–4, 7, 10, 19, 53, 141, 162
Hugo 194
Ike 164, 170, 192–3
Katrina 6, 10, 43, 44–5, 49, 53, 104, 139, 141, 143, 148–51, 161, 189, 191, 193, 205, 206, 209–10, 212
Rita 141, 143
Sandy 6, 8–9, 10, 31–2, 42–6, 47–52, 53, 189, 191–3, 195–8, 200–201, 203, 205–6, 208, 210, 212

industrialized nations 206, 212
insurance 8, 100, 102, 106–8, 111, 118–22, 123, 130, 143, 166, 168, 174, 192, 196, 202, 207, 209
intergovernmental relations 32, 41, 44–5, 48–51
international aid 6, 8, 19, 74, 76

Kobe 9, 35, 37–8, 53, 61–70, 76, 208, 209, 211–12

leadership 8, 9, 52, 76, 83, 207, 211
loss of life 1, 33, 36, 82, 202

military 87, 145
multi-family 154

National Flood Insurance Program 32, 210
National Guard 46, 50, 145
neoliberalism 5, 211–12
nuclear power 38, 40, 42, 47

power 38, 42–3, 45, 47–51, 86–7, 111–12, 114, 196, 198–9, 203, 210
property 1, 9, 25, 33, 67, 69, 84, 104, 106–9, 118, 120–21, 123, 128, 150, 162, 164, 173–4, 192–3
property damage 33, 62, 99–100, 159, 192, 199, 200
property owners 68–9, 109, 120, 144, 190, 203
property value 118, 143, 152, 154, 168

relocation 63–4, 73, 164–5, 173 *see also* evacuation
research methods 10, 22, 146
residential property 67, 100, 130
resilience 1–2, 4–6, 8–10, 15, 16, 27, 67, 109, 110–15, 129–31, 139, 141, 159–60, 166–7, 168–70, 172, 174–5, 179, 205, 209–10

sea wall 35, 41, 206, 210 *see also* flood wall
single family 68, 151
social justice 163, 179
strategy 9, 26, 38, 64, 67, 73, 87, 110, 148, 164, 166, 175, 208

taxation 48, 84, 118, 122–5, 129, 164, 169, 173–4, 197
tourism 10, 126, 159–70, 172–80, 199
transportation disruption 17, 31, 33, 38, 42, 44, 48–9, 51, 71, 77, 87, 161, 192, 196, 198, 203, 206
tsunami 1, 2, 8, 10, 38, 100, 141, 166–7, 205
Great East Japanese 8, 31–6, 40–42, 208
Indian Ocean 6, 162
Lisbon 82–3

UNESCO 88, 90
UNISDR 175, 203

wildfire 166
World Heritage Sites 81, 88

zoning 68, 102–6, 108, 109–10, 116, 130, 144, 150, 164–5, 168–9, 173

For Product Safety Concerns and Information please contact our EU
representative GPSR@taylorandfrancis.com
Taylor & Francis Verlag GmbH, Kaufingerstraße 24, 80331 München, Germany

www.ingramcontent.com/pod-product-compliance
Ingram Content Group UK Ltd.
Pitfield, Milton Keynes, MK11 3LW, UK
UKHW022131190926
475702UK00013B/1743

* 9 7 8 1 1 3 8 5 4 7 2 7 8 *